Oxford Cambridge and RSA Examinations

GCSE Mathematics Homework Book

HIGHER COURSE

SERIES EDITOR BRIAN SEAGER

HOWARD BAXTER, MIKE HANDBURY, JOHN JESKINS, JEAN MATTHEWS

Hodder & Stoughton

A MEMBER OF THE HODDER HEADLINE GROUP

Orders: please contact Bookpoint Ltd, 130 Milton Park, Abingdon, Oxon OX14 4SB.
Telephone: (44) 01235 827720. Fax: (44) 01235 400454. Lines are open from 9.00 – 6.00, Monday to Saturday, with a 24 hour message answering service. Email address: orders@bookpoint.co.uk

British Library Cataloguing in Publication Data
A catalogue record for this title is available from the British Library

ISBN 0 340 846445

First Published 2002
Impression number   10   9   8   7   6   5   4   3   2   1
Year   2007   2006   2005   2004   2003   2002

Copyright © 2002 by Howard Baxter, Mike Handbury, John Jeskins, Jean Matthews and Brian Seager.

All rights reserved. No part of this publication may be reproduced or transmitted in any form or by any means, electronic or mechanical, including photocopy, recording, or any information storage and retrieval system, without permission in writing from the publisher or under licence from the Copyright Licensing Agency Limited. Further details of such licences (for reprographic reproduction) may be obtained from the Copyright Licensing Agency Limited, of 90 Tottenham Court Road, London W1P 9HE.

Typeset by Macmillan India

Printed in Great Britain for Hodder & Stoughton Educational, a division of Hodder Headline Plc, 338 Euston Road, London NW1 3BH by Arrowsmiths.

Contents

Contents

Introduction

The exercises in this book are designed to be used with the Higher Tier of GCSE Mathematics. It is particularly aimed at the OCR Specification A.

Each exercise here matches the exercises in the GCSE Mathematics for OCR Higher Text Book. In the Text Book the exercises are usually in pairs, **a** and **b**. In this Homework Book they all have the letter **c**. Thus, for example, if you had been working on *Random numbers* in class, the exercises would be 21.5a and 21.5b. The corresponding homework exercise is 21.5c.

You will find that these homework exercises are shorter than those in the Text Book but still cover the same mathematics. Some questions are intended to be completed without a calculator, just as in the Text Book. These are shown with a non-calculator icon in the same way. Doing these questions without a calculator is vital preparation for the first GCSE paper.

The double exercises in the Text Book are there to give you extra practice. These homework exercises extend this idea. It is also a smaller book to carry home! If you have understood the topics, you should be able to tackle these exercises confidently as they are no harder than those you have done in class – in fact, in some cases the exercises here may be a little easier. See if you agree. More practice helps to reinforce the ideas you have learned and makes it easier to remember at a later stage.

If, however, you do forget, further help is at hand. As well as the Text Book and this Homework Book there is also a GCSE Mathematics for OCR Higher Revision Book. This is designed for use nearer to your GCSE exam to help you revise and explains the points you do not understand.

Other titles available:

GCSE Mathematics for OCR Higher Revision Book   0340 85614 9
GCSE Mathematics for OCR Higher Text Book   0340 75870 8

Numbers ①

Approximating numbers

1. Write each of the following to 2 decimal places.
   (a) 80.9346   (b) 5.1174   (c) 4.39852   (d) 0.03499   (e) 649.0019

2. Work out each of the following. Give your answer to the accuracy stated.

   (a) $\dfrac{5}{6}$ to 3 decimal places

   (b) 5.5^2 to 1 decimal place

   (c) $\sqrt{24}$ to 2 decimal places

   (d) $\dfrac{64.3074}{2.91}$ to 2 decimal places

   (e) $\dfrac{15}{\pi}$ to 3 decimal places

1. Round each of the following numbers
   (a) 1066
   (b) 23 629
   (c) 8912
   (d) 26 788
   (e) 46 950
      (i) to the nearest 10
      (ii) to the nearest 100
      (iii) to the nearest 1000.

2. Round each of the following numbers to the nearest million.
   (a) 1 500 070
   (b) 5 020 469
   (c) 2 964 720
   (d) 4 199 689
   (e) 876 543

1. Round each of the following numbers correct to the number of significant figures stated.
   (a) 24.86583   (i) to 4 s.f.   (ii) to 2 s.f.
   (b) 0.0083582   (i) to 3 s.f.   (ii) to 2 s.f.
   (c) 4.96848   (i) to 5 s.f.   (ii) to 1 s.f.
   (d) 21.989   (i) to 3 s.f.   (ii) to 2 s.f.
   (e) 0.35602   (i) to 4 s.f.   (ii) to 3 s.f.

2. Round each of the following numbers correct to the number of significant figures stated.
   (a) 28387617   to 5 s.f.   (d) 5990   to 2 s.f.
   (b) 9143   to 3 s.f.   (e) 762.105   to 2 s.f.
   (c) 419.9   to 1 s.f.

3. Work out each of the following. Give your answer correct to the number of significant figures stated.
   (a) 8.42×8.8 to 3 s.f.   (d) 18×0.0703 to 3 s.f.
   (b) $63 \div 2.9$ to 2 s.f.   (e) $0.125 \div 514$ to 4 s.f.
   (c) 59^2 to 2 s.f.

Ordering numbers

1. Write these numbers in order, smallest first.
   3.11, 3.1, 3.01, 3.111, 3.011

2. Write these numbers in order, largest first.
   2.705, 27.5, 2.75, 0.275, 2.075, 27.05

3. Write these numbers in order, smallest first.
   5020, 5220, 5002, 5200, 5202, 5022

Ratio

1. Write the following ratios in their simplest form.
   - (a) $9 : 15$
   - (b) $8 : 12$
   - (c) $32 : 80$
   - (d) $90 : 42$

2. Complete the following ratios.
   - (a) $3 : 4 = 6 : \underline{}$
   - (b) $18 : 9 = \underline{} : 1$
   - (c) $\underline{} : 5 = 12 : 15$
   - (d) $12 : 10 = \underline{} : 1$
   - (e) $\underline{} : 6 : 7 = 15 : 30 : \underline{}$

3. Write these ratios in their simplest form.
   - (a) 40 m : 60 m
   - (b) £2 : 20 p
   - (c) 6 cm : 5 mm
   - (d) 2 kg : 750 g
   - (e) 2 hours : 20 min

4. The ratio of boys to girls in a class is $6 : 4$.
   There are 12 girls in the class. How many boys are there?

5. A map is drawn to a scale of $1 : 50\,000$.
   Two villages are 7 cm apart on the map.
   What is the real distance between them?

Negative numbers

1. Work out the following.
   - (a) -3×6
   - (b) $-8 \div -4$
   - (c) -9×-4
   - (d) $30 \div -2$
   - (e) 4×-7

2. Work out the following.
   - (a) $-3 \times -3 \times -3$
   - (b) $(-12 \div 3) \times -5$
   - (c) $(-9)^2$
   - (d) $\dfrac{-7}{-4 - 3}$
   - (e) $(-3 + 5) \times -2$

3. Given that $x = 3$ and $y = -6$ find the value of:
   - (a) $2y$
   - (b) $2x - y$
   - (c) xy
   - (d) y^2
   - (e) $5y - 4x$

Indices

Exercise 1.7c

1. Express the following in index form.
   (a) $3 \times 3 \times 3 \times 3$ (b) $2 \times 2 \times 2 \times 2 \times 2 \times 2$
   (c) $4 \times 4 \times 5 \times 5 \times 5$ (d) $2 \times 7 \times 7 \times 7 \times 7$
   (e) $\dfrac{1}{5 \times 5 \times 5 \times 5 \times 5}$

2. Simplify the following.
   (a) $3^4 \times 3^2$ (b) $2^5 \times 2^3 \times 2$ (c) $5^8 \div 5^2$
   (d) $4^6 \times 4^{-3}$ (e) $\dfrac{8^5}{8^{-2}}$

3. Work these out and give your answer in index form.
   (a) $\dfrac{2^4 \times 2^3}{2^2}$ (b) $\dfrac{6^7 \times 6^{-3}}{6^2 \times 6^8}$ (c) $\dfrac{3}{3^4 \times 3^2}$

Standard form

Exercise 1.8c

1. Write these numbers in standard form.
   (a) 79 000
   (b) 100
   (c) 0.0034
   (d) 46.3
   (e) 0.000 094 1

2. Write these as ordinary numbers.
   (a) 7.53×10^3
   (b) 8.73×10^{-1}
   (c) 3.192×10^6
   (d) 4.51×10^{-4}
   (e) 6×10^5

Exercise 1.9c

1. Work out these calculations. Give your answers in standard form.
   (a) $(3 \times 10^2) \times (2 \times 10^2)$ (d) $(5.3 \times 10^3) + (8.2 \times 10^3)$
   (b) $(3.6 \times 10^6) \div (9 \times 10^2)$ (e) $(2.7 \times 10^{-3}) \times (5 \times 10^{-2})$
   (c) $(3.3 \times 10^4) - (4.6 \times 10^3)$

Powers and roots

1. Write down the square of each of the following.

    (a) 5   (b) 11   (c) 20

2. Write down the square root of each of the following.

    (a) 64   (b) 144   (c) 625

3. Write down the cube of each of the following.

    (a) 2   (b) 6   (c) 10

4. Write down the cube root of each of the following.

    (a) 27   (b) 125   (c) 512

Reciprocal

1. Write down the reciprocal of each of the following.

    (a) 7   (b) 12   (c) 90

2. Work out the reciprocal of each of the following.

    (a) 2   (b) 5   (c) $\frac{1}{4}$ (d) 0.1

Angle properties

Exterior angle of a triangle

Exercise 2.1c

Calculate the sizes of all the angles marked with letters.

1.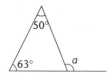

2.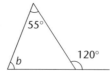

3.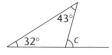

4.

5.

6.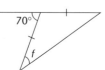

7.

8.

Angle in a semicircle

Exercise 2.2c

In each question, AB is a diameter. Calculate the size of the angles marked with letters. O is the centre of the circle.

1.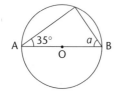

2.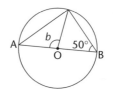

3.

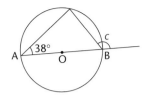

4.

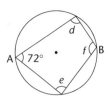

Angles in polygons

1. Calculate the number of sides of the regular polygons with these interior angles.

   (a) 170°   (b) 165°   (c) 108°

2. Calculate the interior angle of the regular polygons with these number of sides.

   (a) 72   (b) 45

Angles in circles

Calculate the sizes of all the angles marked with letters. O is the centre of the circle.

1.

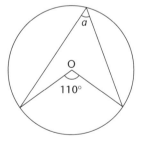

2.

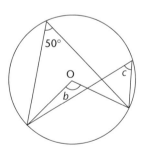

3.

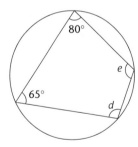

4.

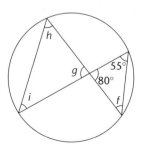

5.

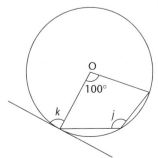

6.

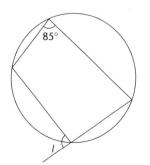

3 Calculating and representing data

Grouped discrete data

1. A group of children counted the number of chocolate drops in some bags of these sweets. These were the results:

Number of chocolate drops	Frequency
29	1
30	3
31	8
32	7
33	4
34	2

(a) How many bags of chocolate drops were counted?

(b) What was the total number of chocolate drops in these bags?

(c) What was the mean number of chocolate drops in these bags?

2. For the data in Question 1, state

(a) the mode   (b) the median

(c) the range.

3. This table shows the number of goals scored by a hockey team in their matches one season.

Goals	0	1	2	3	4	5
Frequency	2	5	6	1	3	1

(a) What was the modal number of goals?

(b) Calculate the mean number of goals scored per match.

Grouped continuous data

1. State the boundaries of these intervals.

(a) 24 cm, to the nearest cm

(b) 6–8 kg, to the nearest kg

2. State the midpoints of these intervals.

(a) $25 \text{ cm} < \text{length} \leqslant 30 \text{ cm}$

(b) 6–8 kg, to the nearest kg

3. Here are the heights of some fig plants.

Height (cm)	Frequency
$20 \leqslant h < 40$	2
$40 \leqslant h < 60$	3
$60 \leqslant h < 80$	8
$80 \leqslant h < 100$	6
$100 \leqslant h < 120$	1

Calculate an estimate of the mean height of these plants.

4. (a) Calculate an estimate of the mean of these masses of packs of cheese.

Mass (g)	Frequency
$100 \leqslant m < 200$	4
$200 \leqslant m < 300$	36
$300 \leqslant m < 400$	47
$400 \leqslant m < 500$	21
$500 \leqslant m < 600$	12

(b) Draw a bar graph to represent this distribution.

5. (a) Draw a frequency polygon for this distribution of the handspan of some students.

Handspan (in cm)	Frequency
$16 \leqslant h < 17$	3
$17 \leqslant h < 18$	4
$18 \leqslant h < 19$	7
$19 \leqslant h < 20$	9
$20 \leqslant h < 21$	4
$21 \leqslant h < 22$	2
$22 \leqslant h < 23$	1

(b) Calculate an estimate of the mean of these handspans.

④ Transformations

Reflections and rotations

Exercise 4.1c

1. Draw axes from −5 to 5.

   Draw the rectangle with vertices (1, 2), (2, 2), (2, 4), (1, 4). Label it P.

   (a) Draw the reflection of P in $y = 1$. Label the image A.

   (b) Draw the reflection of P in $y = -x$. Label the image B.

   (c) Rotate P through 90° anticlockwise about (0, 1). Label the image C.

2. Look at the following diagram.

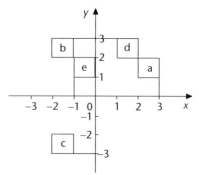

   Describe fully the single transformation that maps:

   (a) a onto b   (b) a onto c

   (c) a onto d   (d) a onto e

   (e) b onto d.

Translations and enlargements

Exercise 4.2c

1. Draw axes from −6 to 6.

   Draw the triangle with vertices (−4, 2), (−3, 2), (−3, 4). Label it P.

   (a) Translate P through the vector $\begin{pmatrix} 5 \\ 1 \end{pmatrix}$. Label the image A.

   (b) Translate P through the vector $\begin{pmatrix} 2 \\ -5 \end{pmatrix}$. Label the image B.

   (c) What translation maps A onto B?

2. Draw axes from −6 to 6.

   Draw the rectangle with vertices (1, 1), (3, 1), (3, 2), (1, 2). Label it Q.

   (a) Enlarge Q with scale factor 2, centre (0, 0).

   (b) Enlarge Q with scale factor 3, centre (4, 0).

3. Look at the following diagram.

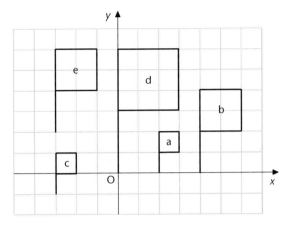

Describe fully the single transformation that maps:

(a) a onto b   (b) a onto c

(c) a onto d   (d) b onto e.

Combining transformations

Exercise 4.3c

In this exercise, carry out the transformations on any object of your choice. In each case describe fully the single transformation that is equivalent.

1. Reflection in $y = 3$ followed by reflection in $y = 4$.

2. Reflection in $y = x$ followed by reflection in $y = -x$.

3. Translation through the vector $\begin{pmatrix} 5 \\ 2 \end{pmatrix}$ followed by translation through the vector $\begin{pmatrix} -2 \\ -5 \end{pmatrix}$.

4. Enlargement scale factor 2, centre O followed by enlargement scale factor 3, centre O.

5. Enlargement scale factor 2, centre O followed by translation through vector $\begin{pmatrix} 2 \\ 0 \end{pmatrix}$.

⑤ Calculations

Fractions

Exercise 5.1c

1. Fill in the blanks in these equivalent fractions.

   (a) $\dfrac{1}{4} = \dfrac{}{8} = \dfrac{}{12} = \dfrac{5}{}$

   (b) $\dfrac{2}{5} = \dfrac{4}{} = \dfrac{}{25} = \dfrac{12}{}$

2. Write these fractions in their lowest terms.

   (a) $\dfrac{6}{9}$ (b) $\dfrac{15}{25}$

   (c) $\dfrac{4}{12}$ (d) $\dfrac{18}{27}$

3. Add these fractions.

   (a) $\dfrac{1}{8} + \dfrac{1}{8}$

   (b) $\dfrac{1}{4} + \dfrac{3}{8}$

   (c) $\dfrac{1}{4} + \dfrac{1}{5}$

   (d) $\dfrac{1}{3} + \dfrac{3}{10}$

   (e) $\dfrac{3}{8} + \dfrac{1}{6}$

4. Subtract these fractions.

   (a) $\dfrac{3}{8} - \dfrac{1}{8}$

   (b) $\dfrac{5}{8} - \dfrac{1}{4}$

   (c) $\dfrac{1}{3} - \dfrac{1}{8}$

   (d) $\dfrac{5}{6} - \dfrac{3}{8}$

   (e) $\dfrac{5}{8} - \dfrac{2}{5}$

Mixed numbers

Exercise 5.2c

1. Change these top-heavy fractions to mixed numbers.

   (a) $\dfrac{11}{8}$ (b) $\dfrac{15}{8}$ (c) $\dfrac{9}{4}$ (d) $\dfrac{7}{2}$ (e) $\dfrac{15}{7}$

2. Change these mixed numbers to top-heavy fractions.

   (a) $1\dfrac{1}{8}$ (b) $2\dfrac{5}{8}$ (c) $3\dfrac{1}{4}$ (d) $5\dfrac{1}{2}$ (e) $5\dfrac{1}{5}$

3. Add these fractions and write your answers as simply as possible.

   (a) $\dfrac{3}{10} + 1\dfrac{2}{5}$ (b) $1\dfrac{1}{4} + \dfrac{3}{5}$ (c) $2\dfrac{1}{5} + 1\dfrac{1}{3}$ (d) $4\dfrac{1}{2} + 2\dfrac{3}{5}$ (e) $1\dfrac{5}{6} + \dfrac{2}{5}$

4. Subtract these fractions and write your answers as simply as possible.

   (a) $2\dfrac{3}{10} - \dfrac{1}{10}$ (b) $3\dfrac{5}{6} - 1\dfrac{3}{8}$ (c) $4\dfrac{1}{4} - 2\dfrac{1}{8}$ (d) $3\dfrac{1}{2} - \dfrac{3}{5}$ (e) $6\dfrac{1}{10} - 4\dfrac{2}{5}$

Multiplying and dividing fractions

Work out the following.

1. (a) $\dfrac{1}{4} \times \dfrac{3}{5}$

    (b) $\dfrac{1}{5} \times \dfrac{1}{4}$

2. (a) $\dfrac{3}{4} \times \dfrac{2}{3}$

    (b) $\dfrac{4}{5} \times \dfrac{1}{2}$

3. (a) $\dfrac{3}{4} \times \dfrac{8}{9}$

    (b) $\dfrac{5}{6} \times \dfrac{3}{5}$

4. (a) $\dfrac{1}{4} \div \dfrac{3}{5}$

    (b) $\dfrac{1}{5} \div \dfrac{3}{5}$

5. (a) $1\dfrac{1}{4} \times \dfrac{3}{5}$

    (b) $2\dfrac{1}{2} \times 1\dfrac{3}{5}$

6. (a) $1\dfrac{1}{4} \div \dfrac{5}{6}$

    (b) $3\dfrac{1}{3} \div 4\dfrac{1}{6}$

Adding and subtracting negative numbers

1. (a) $(-2 \times -4) + (3 \times -2)$

    (b) $(-5 \times 4) - (-2 \times 8)$

2. $-4 + 3 - 2 - 1 + 1 + 3 + 4 - 5$

3. (a) $\dfrac{-5 + 1}{2}$ (b) $\dfrac{4 - 9}{8 - 3}$

Use a calculator to work these out. Give the answers exactly or to 3 significant figures.

4. (a) $-1.43 + 2.79 - 3.64$

    (b) $(-4.1 \times -3.7) + (2.9 \times -3.6)$

5. (a) $\dfrac{4.6 - 3.7}{9 - 7.4}$ (b) $\dfrac{-2.7 \times 3.9}{2.6 + 3.7}$

Work these out. Give the answers exactly or to 3 significant figures.

1. (a) 3.2^4 (b) $310^{\frac{1}{5}}$

2. (a) $\sin 21.4°$ (b) $\cos 48.9°$

    (c) $\tan^{-1} 1.24$ (d) $\cos^{-1} 0.123$

3. (a) $\sin^{-1}\dfrac{4.7}{5.9}$

    (b) $\sin 47° - \cos 47°$

    (c) $\sqrt{2^2 - 1.8^2}$

4. (a) $4.2 \times 10^5 \times 3.9 \times 10^{-1}$

    (b) $2.2 \times 10^{-1} + 3.1 \times 10^{-2}$

5. (a) $2.3 \times 4.7^2 - 4.6 \div 2.89$

    (b) $\dfrac{4.67 \times 3.91 - 4.26}{6.42 - 3.97}$

Ratio and proportion

Exercise 5.6c

1. Split £1950 in the ratio 4 : 5 : 6.

2. Nasim and Andre share a bill in the ratio 2 : 3. Nasim pays £8.40. How much does Andre pay?

3. Sue, Audrey and Aggie invest £70 000 between them in the ratio 2 : 3 : 5.
   How much does each invest?

4. Paint is mixed with 2 parts white to 3 parts black to make grey paint.
   (a) How much black should be mixed with 5 litres white?
   (b) How much black is needed to make 2.5 litres of grey paint?

5. An alloy is made from copper, iron and nickel in the ratio 2 : 3 : 1.
   (a) How much iron is there in 360 g of the alloy?
   (b) How much (i) iron and
       (ii) nickel are needed with 260 g of copper?

Repeated proportional changes

Exercise 5.7c

1. What do you multiply a quantity by if it is increased by
   (a) 2%   (b) 12%   (c) 4.5%
   (d) $\frac{1}{4}$ (e) $\frac{2}{5}$?

2. What do you multiply a quantity by if it is decreased by
   (a) 2%   (b) 12%   (c) 4.5%
   (d) $\frac{1}{4}$ (e) $\frac{2}{5}$?

3. A car priced at £8000 is reduced by 5%. What is the new price?

4. £1000 is invested and 3% is added to the balance at the end of each year.
   How much is in the account after 5 years? Give the answer to the nearest penny.

5. A coat was priced at £816. Each day the price was reduced by $\frac{1}{5}$ until it was sold. It was sold after four reductions. For what price was it sold? Give the answer to the nearest penny.

Finding the value before a percentage change

1. After a 10% increase a man's wage was £297. What was it before the increase?

2. The price of a car after a 5% reduction was £10 450. What was the price before the reduction?

3. The cost of a mobile phone increased by 8%. It is now £88.56. What was it before the increase?

4. In a sale all items were reduced by $\frac{1}{5}$. The sale price of a coat is £64. How much was it before the sale?

5. The cost of a TV including VAT at 17.5% is £176.25. What is the cost excluding VAT?

Equations and manipulation I

Substituting numbers in a formula

Exercise 6.1c

Work out each of the formulae for the values given.

1. $v = u + at$ when $u = 6$, $a = 4$, $t = \frac{1}{2}$

2. $z = 4y - 5x$ when $y = 7$, $x = -2$

3. $k = fh + \dfrac{g}{h}$ when $f = 1.2$, $g = 3.5$, $h = 0.5$

4. $b = a^2 - c^3$ when $a = -5$, $c = -3$

5. $p = \dfrac{a}{2bc}$ when $a = 6$, $b = \frac{3}{4}$, $c = -2$

Collecting like terms and simplifying expressions

Exercise 6.2c

Simplify the following.

1. $x + y + 2x + 5y$

2. $7p + 3q - 2p - 2q$

3. $8a - 2b - 6a - 3b$

4. $3x^2 - 2x + 4x^2 + 5x - 2x^2 - 7x$

5. $3x - 2y + 4z - 2x - 3y + 5z + 6x + 2y - 3z$

6. $1.2x^3 - 3.4x^2 + 2.6x + 3.8x^2 + 3.7x - 2.8$

7. $3(2x + 1) + 2(x + 4)$

8. $5(2y + 3) - 3(4y + 2)$

9. $3(p + 5) - (3p - 4)$

10. $5x(2x - y) + 3x(x + 2y)$

Multiplying out two brackets

Multiply out the brackets.

1. $(x + 4) (x + 5)$
2. $(3x + 7) (x + 6)$
3. $(2x + 4) (3x + 2)$
4. $(x - 6) (x - 3)$
5. $(x + 3) (x - 1)$
6. $(3x - 5) (2x + 3)$
7. $(x - 8) (4x - 1)$
8. $(3x + 7)^2$
9. $(3x - 4y) (2x + 3y)$
10. $(5a + 2b) (5a - 2b)$

Simplifying expressions using indices

Simplify each of the following.

1. $6x^2 \times 3x^4$
2. $\dfrac{10p^8}{5p^5}$
3. $\dfrac{4c^3 \times 7c^5}{2c^2 \times c^3}$
4. $\dfrac{(8y^6)^2 \times 3y}{12y^3}$
5. $\dfrac{3pq \times 16p^3q^2}{12p^7q^3}$

Finding the *n*th term of a sequence

Find the *n*th term of each of these sequences.

(Remember, the formula for the sequence 1, 4, 9, 16, 25, 36, ... is n^2.)

1. 8, 11, 16, 23, 32, 43, ...
2. 2, 8, 18, 32, 50, 72, ...
3. −1, 8, 23, 44, 71, 104, ...
4. 5, 11, 19, 29, 41, 55, ...
5. 6, 20, 44, 78, 122, 176, ...

Factorising algebraic expressions

Factorise.

1. $6x + 3y$
2. $6x^2 - 4x$
3. $3a^2 - ab + ac$
4. $2y + 8xy$

5. $8ab - 4ab^2$
6. $a^3 + a^2b$
7. $6x^3 - 15xy^2$
8. $8x^2yz + 4xy^2z + 12xyz$

Factorising expressions of the type $x^2 + ax + b$

Factorise.

1. $x^2 + 7x + 12$
2. $x^2 - 8x + 15$

3. $a^2 + 14a + 24$
4. $x^2 - 5x + 6$
5. $x^2 - 6x + 9$

6. $x^2 + 10x + 25$
7. $p^2 + 13p + 12$

Factorise.

1. $x^2 + 2x - 15$
2. $x^2 - 3x - 28$

3. $x^2 + 6x - 7$
4. $x^2 - x - 12$
5. $p^2 - 5p - 14$

6. $y^2 - 9$
7. $x^2 + 5x - 36$

Rearranging formulae

Rearrange each formula to make the letter in the brackets the subject.

1. $a + b + c = 180$ (b)
2. $x = 5y + z$ (y)
3. $A = \dfrac{bh}{2}$ (h)

4. $d = \sqrt{15s}$ (s)
5. $a = 180(n - 2)$ (n)
6. $x = ut + at$ (t)

Inequalities

Exercise 6.10c

1. Write down the integer values of x when $-5 \leqslant x < 3$

Solve these inequalities.

2. $4x - 3 > 9$
3. $2(y - 1) < 6$
4. $1 - 3p \geqslant 10$

5. $0 < 2x - 3$
6. $5x - 4 > 3x + 2$
7. $3t + 8 \leqslant 5t - 2$
8. $4y + 13 < 3(2y - 1)$

Forming equations and inequalities

Exercise 6.11c

1. One number is three times another. Their sum is 24.

   (a) Taking x as the smaller of the two numbers, write down an equation in x.

   (b) Solve your equation and find what the two numbers are.

2. I think of a number, x, double it and add 11. The answer is 37.

   (a) Write down an equation in x.

   (b) Solve your equation to find the number I first thought of.

3. A TV repair man charges a call out fee of £40 and £24 per hour for each hour that he works.

   (a) Write down a formula for his total charge when he works x hours.

   (b) For one job he receives £136. Write down an equation in x and solve it to find how many hours the job takes.

4. The diagram below shows a plan of my garden.

   There is a gateway, 3 metres wide. I will need to buy at least x fence panels, each 2 metres wide, to be certain of fencing around the rest of the garden.

   (a) Write down an expression in x for the perimeter of the garden.

   (b) My garden has a *total* perimeter of 56 metres.

   Write down an inequality in x and solve it.

   (c) What is the smallest number of fence panels that I must buy?

Questionnaires and cumulative frequency

Box-and-whisker plots

Exercise 7.1c

1. Amrit did a survey on the times spent on homework in the previous week at her school. The results for the 120 students in year 10 are in the table below.

Time t (hours)	Frequency
$0 \leqslant t < 2$	2
$2 \leqslant t < 4$	11
$4 \leqslant t < 6$	21
$6 \leqslant t < 8$	36
$8 \leqslant t < 10$	29
$10 \leqslant t < 12$	15
$12 \leqslant t < 14$	6

(a) Draw a cumulative frequency graph for year 10 and find the median and quartiles.

(b) Show these on a box-and-whisker plot. Assume a minimum time of 0 hours and a maximum of 14 hours.

(c) The cumulative frequency graph for the 100 students in year 11 is drawn below. Use it to find the median and quartiles for year 11.

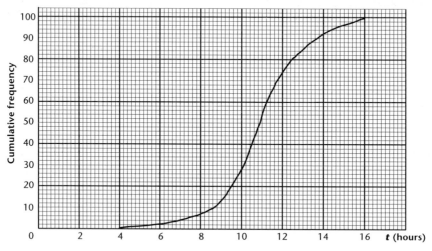

(d) Draw a box-and-whisker plot for year 11.

(e) Compare the results for each year group.

2. Draw a box-and-whisker plot for this set of data.

6 12 10 22 5 17 16 20 9 6 7 24 14 9 19

Solving problems and ⑧ checking results

Checking answers by rounding to one significant figure

Exercise 8.1c

1. Find approximate answers to these calculations by rounding each number to 1 significant figure.

   (a) $61.7 \div 5.8$ (b) 3.7×8.1

   (c) 23.127×28.4

   (d) 73.4×46.8

   Now use a calculator to see how close your approximations are to the correct answers.

2. Find approximate answers to these calculations by rounding each number to 1 significant figure.

   (a) $\dfrac{17.8 \times 5.7}{39.2}$

   (b) $\sqrt{9.7 \times 11.2}$

   (c) 0.82×27.3

   (d) $\dfrac{0.58 \times 73.4}{6.12}$

Compound interest

Exercise 8.2c

1. These accounts earn compound interest. Find the amount in the account for:

   (a) £500 invested at 3% for 2 years

   (b) £250 invested at 5% for 4 years

   (c) £2000 invested at 5.6% for 3 years

   (d) £500 invested at 4.5% for 6 years.

2. £500 is invested at 6% per year compound interest. How long must I leave it before there is £700 in the account?

Insurance

Exercise 8.3c

1. Petra has a 60% no claims bonus. How much car insurance must she pay if the basic premium for her car is £650?

2. This table shows the monthly cost in £ of some health insurance for non-smokers. There is an additional 10% charge for smokers.

Age (years)	Female	Male
18–29	17.99	18.99
30–39	20.99	21.99
40–49	23.99	25.99
50–59	26.99	29.99
60–69	29.99	32.99
70 or over	32.99	35.99

Find the monthly premiums for a

(a) 24 year old woman who does not smoke

(b) 43 year old man who smokes

(c) married couple: wife aged 53 who smokes, husband aged 55, non-smoker.

Compound measures

Exercise 8.4c

1. A car travels 90 miles in 2 hours. What is its average speed?

2. A train is travelling at 22 m/s. How far does it travel in 1 minute?

3. A runner's average speed is 3.2 m/s. How long does it take her to run 1 km?

4. Greenborough has a population of 86 000 and an area of 160 km^2. Calculate its population density.

5. A chocolate has a mass of 30 g and a density of 12 g/cm^3. Calculate its volume.

Working to a reasonable degree of accuracy

Exercise 8.5c

Write down sensible values for each of these measurements.

1. The length of a garden given as 17.284 m.

2. A baby weighing 3 kg 276 g.

Give the answer to the following calculations to a reasonable degree of accuracy.

3. The length of the side of a square table with area 1.7 m^2.

4. The population density of a town with population 120 000 and area 190 km^2.

Pythagoras' theorem and trigonometry

Exercise 9c

1. Calculate the lengths of the missing sides in these right-angled triangles. All lengths are in centimetres.

(a)

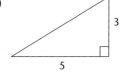

(b)

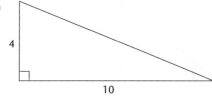

(c)

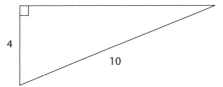

(d)

2. Calculate the lengths of the sides labelled x. All lengths are in centimetres.

(a)

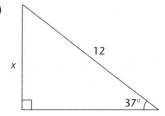

(b)

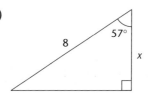

(c)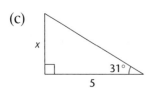

(d)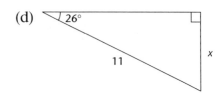

(e)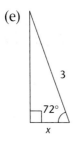

(f)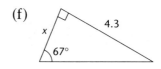

3. Calculate the sizes of the angles labelled x.

(a)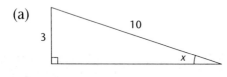

(b)

(c)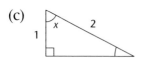

(d)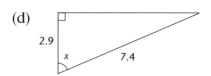

(e)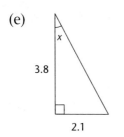

(f)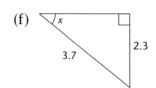

Equations and manipulation 2

Solving harder linear equations

Exercise 10.1c

Solve these equations.

1. $2(x - 3) = x$

2. $3(2x + 1) = 27$

3. $2(2x - 3) = 5(x - 2)$

4. $3(2x - 4) = 2(x - 1)$

5. $\dfrac{x}{2} = 4$

6. $\dfrac{3x}{2} = 5$

7. $\dfrac{5}{x} = 15$

8. $\dfrac{x}{3} = 2x - 5$

9. $2.3x = 9.43$

10. $5.4(x - 3) = 3.78$

Solving inequalities

Exercise 10.2c

Solve these inequalities.

1. $4x < 7 - 3x$

2. $9x - 7 > 1 + 5x$

3. $3(4x - 5) < 9x - 7$

4. $x^2 > 25$

5. $3x - 4 > 4x - 9$

Forming equations and inequalities

Exercise 10.3c

1. A number and twice the number add up to 9. Let the number be x.

   Write down an equation in x and solve it.

2. Three angles of a triangle are x, $2x - 40$ and 70.

   Write down an equation in x and solve it.

3. A number add 7 is the same as twice the number minus 8. Let the number be x.

   Write down an equation in x and solve it.

4. A rectangle has a length of x and a width of $2x - 9$. The perimeter of the rectangle is 12 cm.

   Write down an equation in x and solve it.

5. The square of a number minus 7 is less than 42. Set up an inequality and find the possible values of the number.

Simultaneous equations

Solve these simultaneous equations.

1. $x + y = 9$
    $2x - y = 3$

2. $2x + y = 4$
    $4x + y = 6$

3. $3x + 2y = 9$
    $3x + y = 3$

4. $3x + y = 7$
    $2x - y = 3$

5. $3x + 2y = 15$
    $3x - y = 6$

Solve the simultaneous equations.

1. $x + 2y = 5$
    $2x - y = 0$

2. $2x + y = 7$
    $3x + 2y = 12$

3. $3x + 2y = 9$
    $4x + y = 7$

4. $3x + y = 17$
    $2x - 3y = 4$

5. $3x + 2y = 7$
    $5x - y = 16$

Solving quadratic equations

Solve these equations by factorisation.

1. $x^2 - 4x + 3 = 0$

2. $x^2 + 6x + 5 = 0$

3. $x^2 - 4x - 12 = 0$

4. $x^2 - 8x + 15 = 0$

5. $x^2 - 3x - 10 = 0$

6. $x^2 - 9x + 20 = 0$

7. $x^2 - 7x - 8 = 0$

8. $x^2 - 4x - 21 = 0$

9. $x^2 + 5x + 6 = 0$

10. $x^2 - x - 30 = 0$

Graphical methods of solving equations

Solving graphically the following simultaneous equations.

1. $y = 2x - 3$, $y = 9 - x$. Use values of x from 0 to 6.

2. $y = 3x - 6$, $y = 6 - x$. Use values of x from 0 to 6.

3. $y = 2x$, $2y = 6 + x$. Use values of x from 0 to 6.

4. (a) Draw the graph of $y = x^2 - 5x + 6$ for values of x from 0 to 5.

    (b) Solve the equation $x^2 - 5x + 6 = 0$ from the graph.

5. (a) Draw the graph of $y = x^2 - 2x - 8$ for values of x from -3 to 5.

    (b) Solve the equation $x^2 - 2x - 8 = 0$ from the graph.

Solving cubic equations by trial and improvement

Exercise 10.8c

Use trial and improvement to find these solutions.

1. A solution of $x^3 = 24$ lies between 2 and 3. Find it correct to 1 decimal place.

2. A solution of $x^2 + 5x = 30$ lies between 3 and 4. Find it correct to 1 decimal place.

3. A solution of $x^3 - 2x = 39$ lies between 3 and 4. Find it correct to 1 decimal place.

4. A solution of $x^3 - x^2 = 35$ lies between 3 and 4. Find it correct to 1 decimal place.

5. A solution of $x^3 - 5x = 0$ lies between -3 and -2. Find it correct to 1 decimal place.

Problems that lead to simultaneous or quadratic equations

Exercise 10.9c

1. The sum of two numbers x and y is 9 and the difference between them is 1.

    Write down two simultaneous equations and solve them to find the two numbers.

2. The square of a number, x, plus three times the number is 10.

    Write down a quadratic equation and solve it to find the values of x.

3. To enter Hodsock priory it costs £a for an adult and £b for a child. For two adults and one child it cost £17, and for one adult and three children it cost £18.50.

    Write down two simultaneous equations and solve them to find the cost of entry.

4. Beryl is 38 years older than her daughter Cathy. The product of their ages is 423.

    Let Cathy's age be x. Write down a quadratic equation and solve it to find Cathy's age.

Showing regions on graphs

1. Write down the inequality that describes the region shaded. For each of the remaining questions draw a grid with axes labelled -5 to $+5$ on both axes.

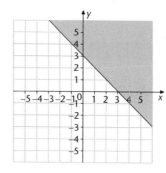

2. Shade the region $x > -2$

3. Shade the region $y < 2x + 1$

4. Shade the region $2y < 4 - x$

Rearranging formulae

For each question make the letter in the brackets the subject.

1. $P = 2a - b$ (a)

2. $R + 2as = 4a + 3t$ (a)

3. $ab - c = ac - 2b$ (a)

4. $R = \dfrac{x}{b} + 3x$ (x)

5. $s - 3x = 4a(b - s)$ (s)

6. $b = a + c^2$ (c)

Measurement and compound units ⓘ

Estimating measurements

1. Estimate the length of this line.

   _____

2. Estimate the size of this angle.

3. Estimate the mass of a house brick.

4. Estimate the amount of liquid in a mug of coffee.

5. Are the following statements reasonable?

   If they are not, then give a more sensible statement.

   (a) A doorway is 2 m high.
   (b) A car's petrol tank holds 30 ml of liquid.
   (c) Your school bag weighs 50 g.
   (d) My stride is 30 ft long.
   (e) It takes Beverley 10 minutes to walk 1 km to school.

Discrete and continuous measures

1. Which of the following is discrete and which is continuous data?

   (a) the amount of sugar in a bag
   (b) the number of students in a class
   (c) the time it takes for you to complete your homework
   (d) the number of marks in a test
   (e) the area of your classroom
   (f) the amount of milk left in the container in the fridge
   (g) your shoe size
   (h) your height

2. To what degree of accuracy would you measure the following things?

   (a) the amount of water in a bucket
   (b) the thickness of an exercise book
   (c) the distance from Hull to York
   (d) the mass of a pencil
   (e) the time to run 20 miles
   (f) the length of a basketball court
   (g) the liquid in a cup
   (h) the weight of a bus
   (i) the diameter of a CD

Bounds of measurement

What is the upper and lower bound of each of the following measures?

1. A pencil, 8 cm long to the nearest cm.

2. A can of cola, 330 ml to the nearest ml.

3. A garden, 14.8 m long to the nearest 0.1 m.

4. A football crowd, 25 000 to the nearest 1000.

5. Time to run 100 m, 9.79 s to the nearest one hundredth of a second.

6. A packet of sweets, 150 g to the nearest 10 g.

Compound units

1. A car travels at an average speed of 65 miles per hour for 3 hours.

   How far was the journey?

2. An 80 cm^3 block of lead has a mass of 880 g.

   What is the density of lead?

3. A train travels 6 km in 4 minutes.

   What is the speed in km/h?

4. A tank holds 1000 litres of water. It is emptied at a rate of 4 l/min.

   How long does it take to empty?

5. The population density of a small town is 50 people per square kilometre.

   There are 2600 people in the town altogether.

   What area does the town cover?

Interpreting graphs

Story graphs

Exercise 12.1c

1. John and Imran live in the same block of flats and go to the same school. The graphs represent their journeys home from school.

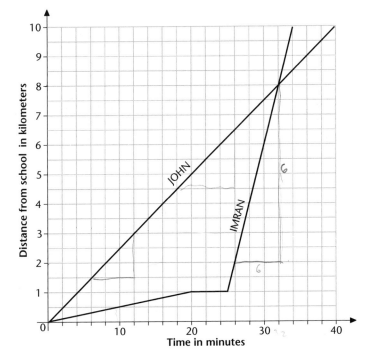

(a) Describe Imran's journey home.

(b) After how many minutes did Imran overtake John?

(c) By how many minutes did Imran beat John home?

(d) Calculate John's speed in kilometres per minute.

2. The graph shows the temperatures at given times on a certain day.

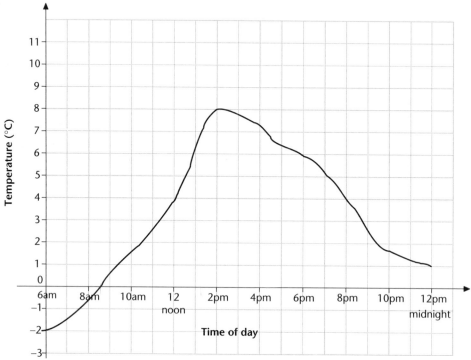

(a) At what time did the temperature first go above freezing point?
(b) What was the maximum temperature?
(c) At what time was the temperature rising most quickly?
(d) For how many hours was the temperature above 6°C?

3. This graph shows the cost of electricity for a given number of units used.

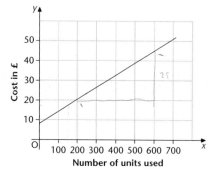

(a) Mrs Jones used 150 units of electricity. What was the cost?
(b) The Green family's bill was £45. How much electricity did they use?
(c) What was: (i) the basic charge (even if no electricity is used)
(ii) the price per unit?

4. Water is poured in each of these vessels at a steady rate.

 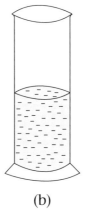 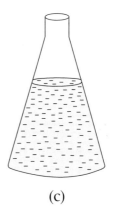

(a)   (b)   (c)

Below are the graphs of depth against time for each of the vessels. State which of graphs (i), (ii) or (iii) best fits each vessel.

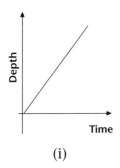 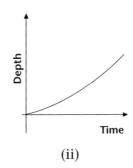 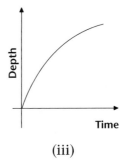

(i)   (ii)   (iii)

Gradient

Exercise 12.2c

1. Find the gradient of each of these lines

(a)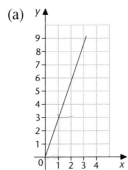

(b)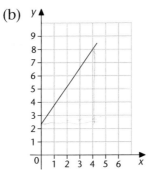

(c)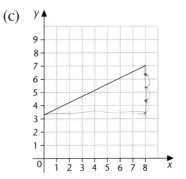

(d)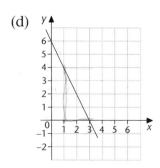

(e)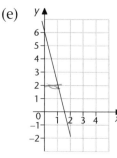

(f)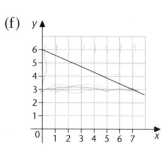

2. Calculate the gradients of the lines joining each of these pairs of points.

   (a) (3, 2) and (5, 14)

   (b) (0, 1) and (2, 9)

   (c) (1, 2) and (3, 5)

   (d) (1, 6) and (5, 2)

   (e) (−1, 2) and (3, 9)

   (f) (−2, 7) and (2, −1)

   (g) (−3, 0) and (5, −2)

   (h) (−3, 5) and (6, 5)

3. Jasvindar drove to visit her grandmother who lives 200 km away.

   The graph shows her journey.

   Find the gradients of the two sloping parts of the graph. What information does this give?

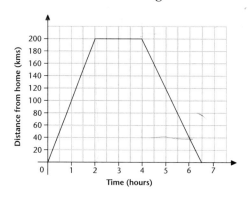

4. Draw, on the same diagram, the graphs of:

   (a) $y = -2x$

   (b) $y = -2x + 3$

   What do you notice? What are the gradients of these lines?

5. Draw a graph for each of these straight lines and find their gradients.

   (a) $y = 3x - 2$

   (b) $y = \frac{1}{2}x + 1$

   (c) $y = -2x + 7$

Straight-line graphs

Exercise 12.3c

1. Write down the equations of the straight lines:

   (a) with gradient 2 and passing through (0, 4)

   (b) with gradient 2.5 and passing through (0, −1)

   (c) with gradient −3 and passing through (0, 10).

2. Look again at Exercise 12.2c, Question 1.

   Use the gradients you found to write down the equations of each of the lines.

3. Find the equations of these lines.

   (a)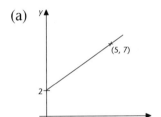

   (c)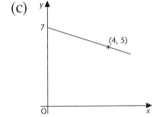

   (b)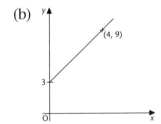

   (d)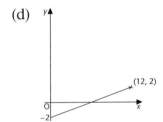

4. Find the gradients of these lines and where they cut the y-axis.

   (a) $y = 4x + 2$ (d) $y + 3 = 2x$

   (b) $y = 5 − 3x$ (e) $y − 5x = 4$

   (c) $y = \frac{1}{2}x − 5$ (f) $2x + 3y = 6$

5. Sketch the graphs of

   (a) $y = 3x + 2$ (b) $y = 5 − 2x$

   (c) $y = \dfrac{12}{x}$ (d) $y = −x^3$.

13 Probability

Covering all the possibilities

1. Gill and Barbie are discussing how to spend the evening. They want to read (R), watch TV (T) or go out (G). They can choose the same or differently. Complete a list to show their possible choices.

2. In a game, two ordinary dice are thrown and the differences between the numbers on the dice is the score. Draw a grid with axes numbered 1 to 6 and show the possible outcomes of the score. What is the probability that the score is 2?

3. **First course**   **Second course**

   Soup   Spaghetti Bolognese

   Melon   Lamb Biriani

                 Chicken and Mushroom Pie

   Copy and complete this tree diagram to show the possible choices.

   First course   **Second course**

   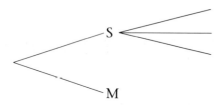

4. A bag contains some counters: red, black or green. Jamal picks out a counter without looking, puts it back, then picks out a second counter. Make a tree diagram to show the possible outcomes.

Probability of event A and event B happening

Exercise 13.2c

1. Kay has 3 black pens, 5 blue pens and 2 red pens in her bag. If she selects a pen at random, what is the probability she chooses a red or blue one?

2. A bag contains 20 balls. There are 6 red ones and 5 yellow ones.

   (a) If a ball is chosen at random, what is the probability that it is not red or yellow?

   (b) Irim chooses a ball at random, puts it back, then chooses again. What is the probability that she has a red ball and then a yellow ball?

3. **First course**   **Second course**

   Soup (0.6)   Spaghetti Bolognese (0.1)

   Melon (0.4)   Lamb Biriani (0.7)

                    Chicken and Mushroom Pie (0.2)

   The numbers in the menu show the probabilities that Rahid chooses these dishes. What is the probability that he chooses Soup and Lamb Biriani?

4. An ordinary die is thrown and a coin is tossed. What is the probability of getting a 6 and a tail?

Using tree diagrams for unequal probabilities

Exercise 13.3c

1. On any day, the probability that Jo cycles to school is 0.8. Copy and complete this diagram to show her possible choices and their probabilities.

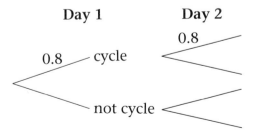

   (a) What is the probability that she does not cycle on either day?

   (b) Find the probability that she cycles on just one of these days.

2. From experience, the probability of Sam winning a Solitaire game is 0.7. Copy and complete this tree diagram to show the probabilities of the outcomes for two games.

First game   **Second game**

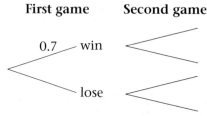

(a) What is the probability that Sam wins the first game but loses the second?

(b) What is the probability that he loses both games?

3. On her way to work, the probability that Rani has to stop at the traffic lights is 0.4. The probability that she has to stop at the level crossing is 0.3. These probabilities are independent.

(a) Complete this tree diagram.

Traffic lights   **Level crossing**

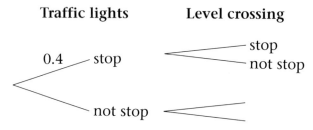

(b) What is the probability that she does not have to stop at all?

(c) Rani thinks that the probability she has to stop just once is 0.28. Show why she is wrong.

Length, area and volume

Area of a parallelogram

Exercise 14.1c

1. Find the areas of these parallelograms. The lengths are in centimetres.

(a)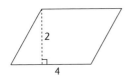

(b)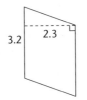

(c)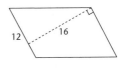

2. Work out the missing lengths (in centimetres).

(a)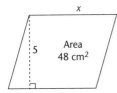

(b)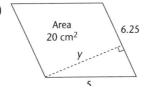

(c)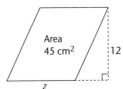

3. The rectangle, parallelogram and triangle all have the same area. Work out the missing lengths.

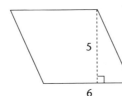

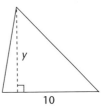

Area of a trapezium

Exercise 14.2c

1. Find the areas of these trapezia. The lengths are in centimetres.

(a)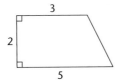

(b)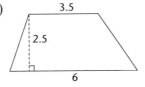

(c)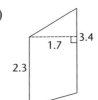

2. Work out the missing lengths in these trapezia (in centimetres).

(a)
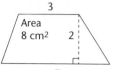

(b)
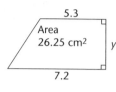

(c)
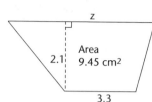

Volume of a prism

1. Calculate the volumes of these.

   (a) A cylinder, radius 2 cm and height 5 cm.

   (b) A prism 30 cm long, with cross section a triangle, base 6.5 cm and height 4.9 cm.

   (c) A prism 45.7 cm long, with cross section a parallelogram, base 12.7 cm and height 5.8 cm.

   (d) A trough of length 250 cm and cross sectional area 40 cm².

2. The volume of a cylinder is 1000 cm³. Its radius is 10 cm. What is its height?

3. The volume of a cylinder is 250 cm³ and its length is 12.3 cm. What is its radius?

Dimensions

All letters, other than π, represent lengths.

State whether these expressions represent length, areas, volumes or none of these.

1. $a^2b + b^2a$

2. $\pi r^2 + \pi rh$

3. $a(b^2 + c)$

4. πr

5. $a(b + c)$

6. $(a^2 + b^2)^2$

7. $\pi r^2h + 4\pi r$

8. $r^3 - 4\pi r^2h$

9. $x(y + 2)$

10. $4\pi r^2 + 2\pi rh$

Properties of transformations and similar shapes

Properties of transformations

Exercise 15.1c

1. Draw a grid with both axes marked from −6 to +6.

   (a) Draw the triangle with vertices at (1, 2), (2, 5) and (3, 5). Label it A.

   (b) Reflect triangle A in the *y*-axis. Label it B.

   (c) Rotate triangle A through 90° clockwise about the origin. Label it C.

   (d) Reflect triangle A in the line *y = x*. Label it D.

   (e) Describe fully the transformation that will map
   (i) D onto C and
   (ii) B onto D.

2. Two triangles both have angles of 70°, 50° and 60°.

   What other information is needed before you can say they are congruent?

3. (a) Find the interior angle of a 12-sided regular polygon.

   (b) Use a diagram to show that a 12-sided regular polygon will not tessellate.

Enlargements

Exercise 15.2c

1. A scale model is made of a house using a scale of $\frac{1}{20}$.

   (a) The height of the real house is 5.8 m. What is the height of the model?

   (b) The length of the bathroom in the model is 9.6 cm. What is the length of the real bathroom?

2. A map is drawn to the scale of 1 : 50 000.

   (a) Two towns are 5 km apart. How far, in centimetres, are they apart on the map?

   (b) The distance between two other towns is 25 cm on the map. What is the real distance between the towns?

3. The triangles ABC and DEF are similar.

   Calculate the lengths of
   (a) AB (b) EF.

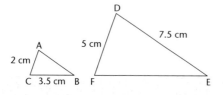

4. This diagram is of Brenda's patio.

   Make a scale drawing using a scale of 2 cm to 1 m.

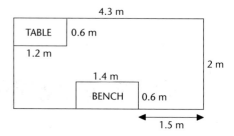

Comparing 16

Comparing data

1. The following marks were obtained by two students in 12 tests.

   Alice   6, 7, 9, 9, 10, 13, 18, 20, 21, 21, 22, 24

   Bob   3, 11, 12, 13, 14, 15, 16, 16, 17, 17, 18, 28

   (a) Find the median and the range of each set of marks.

   (b) Make two comparisons of the two sets of marks.

2. The hours of sunshine recorded at two seaside resorts are listed below.

   Seashore   8.6, 6.4, 12.3, 10.7, 9.5, 7.2, 8.8, 8.2

   Clifftop   2.4, 0, 0, 5.6, 3.2, 9.8, 1.6, 0.4

   (a) Calculate the mean and range of the two sets of data.

   (b) Compare the hours of sunshine at the two resorts.

3. The boxplot below shows the distribution of the weekly wages of men and women who work at a certain factory.

   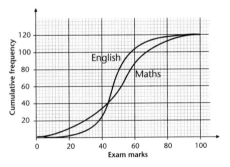

   (a) Find the median and inter-quartile range for the men and the women.

   (b) Compare the wages of men and women at the factory.

4. 120 students took a Maths and an English exam. The cumulative frequency curves below show the distribution of the two sets of marks.

   (a) Use the graph to find the median and the inter-quartile range of the two sets of marks.

   (b) Use your results to compare the two sets of marks.

Correlation

1. The engine size, in cc, and the time to accelerate to 60 mph, in seconds, for a particular range of cars is given in the table below.

Engine size (cc)	1000	1100	1300	1600	1800	1900	2000
Time (s)	15.9	13.9	11.4	10.7	7.0	8.1	6.8

(a) Draw a scattergraph to show this information.

(b) Describe the relationship between the engine size and the time to accelerate to 60 mph.

(c) What type of correlation does this diagram indicate?

(d) Draw a line of best fit onto your graph.

(e) Use your line of best fit to estimate the acceleration time for a car with engine size of 1500 cc.

2. Ten students did a mock exam in November and then another the following March. The results of the two exams are shown in the table below.

Student	A	B	C	D	E	F	G	H	I	J
November exam	32	51	47	35	55	41	63	72	60	45
March exam	37	53	50	38	50	45	64	70	70	50

(a) Draw a scattergraph to show this information.

(b) Describe the relationship between the two sets of scores.

(c) What type of correlation does this diagram indicate?

(d) Draw a line of best fit onto your graph.

(e) Another student obtained 60 marks in the November exam but did not sit the March exam. Use your line of best fit to estimate this student's score for the March exam.

Times series

1. The following table shows the profit (in tens of thousands of pounds) of a shop in each quarter of three successive years.

Year	1st quarter	2nd quarter	3rd quarter	4th quarter
1	13	22	58	23
2	13	28	61	25
3	17	29	61	26

(a) Draw a graph to illustrate this data.

(b) Calculate the four-quarter moving averages.

(c) Plot the moving averages onto your graph.

(d) Comment on the general trend and the quarterly variation.

2. The table below shows the number of guests staying in a small hotel during each month of one particular year.

J	F	M	A	M	J	J	A	S	O	N	D
105	84	99	105	109	114	107	104	115	120	123	115

(a) Draw a graph to illustrate this data.

(b) Calculate the three-month moving averages.

(c) Plot the moving averages onto your graph.

(d) Comment on the general trend and the monthly variation.

⑰ Locus

Identifying a locus

1. Draw a line 8 cm long. Draw the locus of points which are 2 cm from the line.

2. Draw an angle of 110°. Construct the bisector of the angle.

3. Construct a triangle with sides 9 cm, 8 cm and 6 cm. Construct the bisectors of each of the three angles. What do you notice?

4. Draw a rectangle ABCD with sides AB = 8 cm and BC = 5 cm. Show by shading the locus of points, inside the rectangle, which are nearer BA than BC.

5. Construct a triangle ABC with AB = 10 cm, BC = 8 cm and AC = 5 cm. Show by shading the locus of points which are inside the triangle and nearer A than B.

Problems involving intersection of loci

1. Draw a triangle ABC with AB = 8 cm, AC = 6 cm and angle A = 70°. Find, by construction, the point which is equidistant from B and C and is also 5 cm from C. Label the point D.

2. Copy the diagram. The point O is 5 cm from the line AB.

Show, by shading, the locus of points which are less than 3 cm from the line AB and also less than 4 cm from the point O.

3.

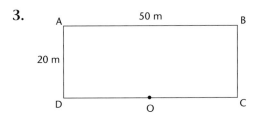

The diagram shows a swimming pool 50 m long by 20 m wide. The shallow end extends to a distance 20 m from AC. Amanda's mother is at O, half way along CD.

(a) Using a scale of 1 cm to 5 m, draw a scale drawing of the pool.

(b) Amanda's mother says Amanda can swim anywhere in the shallow end or within 15 m of her. Show by shading all the places that Amanda can swim.

4. Construct a triangle ABC with AB = 11 cm, AC = 7 cm and BC = 9 cm. Show, by construction, the locus of points, inside the triangle, which are equidistant from A and B and also nearer AB than AC.

5. Draw a rectangle ABCD with AB = 10 cm and AD = 6 cm.

Show by construction the locus of points inside the rectangle for which *all* these statements are true.

(a) The points are nearer to D than A.

(b) The points are nearer DC than DA.

(c) The points are less than 7 cm from A.

18 Proportion and variation

Describe the variation in each of these, using the symbol $\propto$.

1. The volume of water, y litres, in a water butt and the number of times, x, a watering can may be filled from it.

2. The time it takes, t hours, to count the votes after an election and the number of people, y, who are counting.

Describe the variation shown in each of these tables of values. Use the symbol $\propto$.

3.
x	4	12
y	1	3

4.
x	4	2
y	10	20

5.
x	6	3
y	5	10

Variation as a formula

Find formulae for the variations in Exercise 18.1c, Questions 3, 4 and 5.

Other Variation

1. $y \propto x^2$, $y = 10$ when $x = 2$.
   Find y when $x = 4$.

2. $y \propto x^2$, $y = 300$ when $x = 5$.
   Find y when $x = 2$.

3. $y \propto x^3$, $y = 108$ when $x = 6$.
   Find y when $x = 2$.

4. $y \propto x^3$, $y = 24$ when $x = 2$.
   Find y when $x = 20$.

5. $y \propto \frac{1}{x^2}$, $y = 3$ when $x = 2$.
   Find y when $x = 6$.

6. $y \propto \frac{1}{x^2}$, $y = 5$ when $x = 2$.
   Find y when $x = 5$.

Describe the variation shown in each of these tables of values. Use the symbol $\propto$.

7.
x	5	10
y	12.5	100

8.
x	2	3
y	9	4

9.
x	4	2
y	200	50

10.
x	10	2
y	5	125

Formulae

Find formulae for each of the Questions in Exercise 18.3c.

Equations and manipulation 3

Factorising quadratic expressions where the coefficient of $x^2 \neq 1$

Exercise 19.1c

Factorise the following.

1. $x^2 + 8x + 12$
2. $2x^2 - 5x + 2$
3. $4x^2 - 8x + 3$
4. $4x^2 + 21x + 5$
5. $5x^2 - 16x + 3$

Exercise 19.2c

Factorise the following.

1. $x^2 - x - 12$
2. $2x^2 - 5x - 12$
3. $6x^2 + x - 1$
4. $8x^2 - 2x - 3$
5. $6x^2 - 7x - 10$

Solving quadratic equations by the formula

Exercise 19.3c

Solve the equations in Questions 1 to 4 by factorising.

1. $x^2 + x - 6 = 0$
2. $2x^2 - 7x + 5 = 0$
3. $4x^2 + 8x - 5 = 0$
4. $6x^2 + 7x - 3 = 0$

Use the formula to solve the equations in Questions 5 to 8.
Give the answers correct to 2 d.p.

5. $x^2 + 7x + 3 = 0$
6. $2x^2 - x - 4 = 0$
7. $5x^2 + 11x + 1 = 0$
8. $3x^2 - x - 8 = 0$

Algebraic fractions

Simplify.

1. $\dfrac{x}{2} - \dfrac{3x}{5}$

2. $\dfrac{x-1}{3} - \dfrac{4-x}{2}$

3. $\dfrac{1}{x+1} + \dfrac{1}{x-1}$

4. $\dfrac{x}{x+1} - \dfrac{1}{x+2}$

5. $\dfrac{x+1}{x+2} + \dfrac{x+2}{x+1}$

6. $\dfrac{4a^2}{3b^3} \times \dfrac{9b}{(2a)^2}$

7. $\dfrac{3x^2 - 12}{2x^2 - x - 6}$

8. $\dfrac{1}{x} - \dfrac{1}{x-1} + \dfrac{1}{x+1}$

Harder equations and inequalities

Solve.

1. $\dfrac{x}{2} - \dfrac{3x}{8} = \dfrac{5}{12}$

2. $\dfrac{1}{x} - \dfrac{2}{x-2} = 0$

3. $(x+2)(x-5) = 8$

4. $\dfrac{3}{x} + 2x = 7$

5. $\dfrac{1}{x+1} + \dfrac{2}{x+2} = 4$

6. $7 - 2(x-3) < 5 - x$

7. $-5 < 2x - 3 < 6$

8. $x^2 - 2x - 8 < 0$

Rearranging formulae

For each question rearrange to make the letter in square brackets the subject.

1. $p = 2a + at$ [a]

2. $q = b + \dfrac{c}{a}$ [a]

3. $r = \dfrac{a}{a+b}$ [a]

4. $b^2 = c^2 - a^2$ [a]

5. $\dfrac{1}{c} = \dfrac{1}{a} + \dfrac{1}{b}$ [a]

Using graphs to solve harder equations and inequalitites

Do not draw the graphs in Questions 1 and 2.

1. The intersection of two graphs is the solution to the equation $x^2 - 2x - 3 = 0$.

   One of the graphs is $y = x^2 - x - 1$. What is the other graph?

2. The graphs of $y = x^2 + 2x$ and $y = 2x + 1$ are drawn on the same grid.

   What is the equation whose solution is found at the intersection of the two graphs?

3. (a) Draw the graphs of $y = x^2$ and $y = 5 - \frac{1}{2}x$ for $x = -3$ to 3.

   (b) What is the equation of the points where they intersect?

   (c) Solve this equation from your graph.

4. (a) Draw the graph of $y = x^3 - 4x$ for $x = -3$ to 3.

   (i) Draw another graph so that the equation of their points of intersection is $x^3 - 5x + 3 = 0$.

   (ii) Use the graph to solve the equation.

5. (a) Draw the graph of $y = x^2 - 3x$ from $x = -2$ to 5.

   (b) Find the solutions of the following from the graph.

   (i) $x^2 - 3x - 2 = 0$

   (ii) $x^2 - 4x + 1 = 0$

Transformations and congruence

Rotation

Exercise 20.1c

You will need graph paper for this exercise.

1. Draw the triangle ABC with vertices A(1, 3), B(2, 3) and C(4, 1). Show the image following the given rotation.

    (a) Rotate the triangle ABC through 90° clockwise about the origin.

    (b) Rotate the triangle ABC through 180° about the point (1, 2).

2. In each of the following draw the image of the object after a 90° rotation clockwise about O.

    (a) (b) (c)

Exercise 20.2c

You will need graph paper for this exercise.
Draw a set of axes from 0 to 10 for x and y.

1. Draw a triangle ABC with vertices A(4, 1), B(7, 1) and C(7, 5).
   Rotate this triangle 30° about the origin.

2. Draw the square ABCD with vertices A(4, 2), B(4, 4), C(6, 4) and D(6, 2).
   Rotate the square 45° about the point (1, 2).

Exercise 20.3c

You will need graph paper for this exercise.

1. Draw a set of axes from −5 to 5 for the x-axis and 0 to 5 for the y-axis.
   Draw the triangle ABC with vertices A(1, 3), B(1, 5) and C(4, 3).
   Draw another triangle EFG with vertices E(−3, 1), F(−5, 1) and G(−3, 4).
   What rotation will map ABC onto EFG?

2. Draw a set of axes from −6 to 6 for *x* and *y*. Draw the triangle ABC with vertices A(1, 1), B(3, 1) and C(3, 4). Rotate this triangle through 90° about the point (−2, 1) to give the image A′B′C′. Rotate triangle A′B′C′ through 90° about the point (2, 3) to give the image A″ B″ C″. What rotation will map A″ B″ C″ onto the original triangle ABC?

Exercise 20.4c

You will need graph paper for this exercise.

1. Draw a set of axes from −6 to 5 for *x* and *y*. Draw the triangle ABC with the coordinates A(1, 1), B(1, 2) and C(3, 2). Draw the enlargement by a scale factor of −2; the centre of enlargement is the origin.

2. Draw a set of axes from −6 to 6 for *x* and *y*. Draw the triangle ABC with the coordinates A(5, 1), B(5, 4) and C(6, 4). Draw the enlargement by a scale factor of −3; the centre of enlargement is the point (3, 2).

Congruent triangles

Exercise 20.5c

1. Which of these pairs of triangles are congruent?

   Explain your answer in each case. The diagrams are not to scale, but lengths and/or angles are marked.

   (a)

   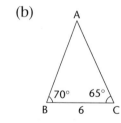

   (b)

   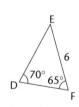

   (c)

   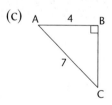
   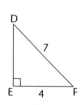

   (d)

   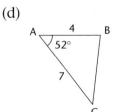
   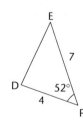

2. Use congruent triangles to prove that the diagonals of a parallelogram bisect each other.

Surveys and sampling ㉑

Surveys

1. You wish to investigate how regularly students at your school visit the cinema.

   You decide to write a questionnaire to find out.

   Write down two or more useful questions that would help you to gather your information.

2. A new local radio station is about to start broadcasting. They decide to send out a questionnaire to find out what sort of programmes are required by their listeners.

   Comment on each of the questions and, where necessary, write a more appropriate one.

   (a) What type of programme do you like most?

   (b) Do you listen to the radio in the afternoon?

   (c) Do you like competitions and phone-ins?

   (d) How much do you earn?

   Are there any other questions that you think will be useful for the radio station?

Sampling

Exercise 21.2c

1. You need to obtain a representative sample of 1000 people for an investigation into how often people eat out at restaurants.

   Comment on the following methods for obtaining the sample:

   (i) By choosing 1000 names from the telephone directory.

   (ii) By stopping 1000 people at random outside the railway station.

   (iii) By asking 100 restaurants to supply 10 names each.

Exercise 21.3c

1. Comment on the following ideas for obtaining a random sample.

   (a) A newspaper editor wishes to judge the public's reaction to the building of a nuclear power station.

   Method: He invites his readers to write to him expressing their views. Then he randomly chooses 10 from the post bag.

   (b) A local council wish to know how often people use public transport.

   Method: They visit every 10th house in each street in the district. If there is no reply they call at the 11th house and so on.

Bias and stratified random sampling

Exercise 21.4c

1. You are asked to conduct an investigation to find out how much time Year 7 students spend watching TV.
   In your school in Year 7 there are 5 groups, each of 30 students.

   (a) How would you obtain a stratified sample of 10% of Year 7?

   In each group there are 40% boys and 60% girls.

   (b) How will this affect the sample that you choose?

2. In a large company there are 4 Departments. There are 175 employees in Department A, 50 employees in Department B, 250 employees in Department C and 125 employees in Department D.

   The Directors wish to consult the employees about a proposal to change the number of hours a week they work. They decide to select a sample of 50 to interview.

   How would you select a stratified sample, so that workers from all Departments are fairly represented?

3. To monitor the number of birds of a particular species, 100 are trapped and tagged. The next year, a sample of 60 birds of the same species are caught. 24 of them are found to be tagged.

   Calculate an estimate of the size of the population of this species of bird.

Random numbers

1. An ornithologist wishes to know the number of birds' nests on a cliff face. It is not realistic to try to count every single nest, so he decides to estimate the number. He divides the area into 36 squares and decides to pick 9 squares at random as his sample.

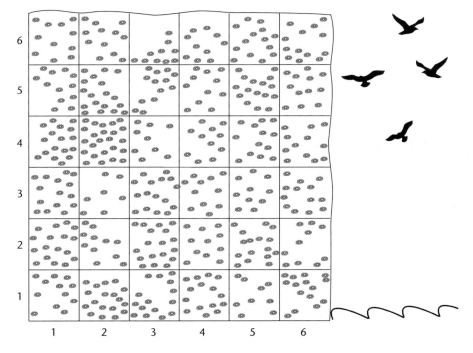

Method: (i) Throw a dice twice to select a square. The first throw gives the 'across' square and the second throw gives the 'up' square.

(ii) Count the number of nests in the chosen square.

(iii) Repeat the process to get the number of nests in 9 *different* squares.

(iv) Multiply your total by 4 to get an estimate of the total number of nests.

Try this two or three times. How do your answers compare? Ask others in the class what answers they got. How do they compare with the exact total number of nests?

Indices

Fractional indices

Exercise 22.1c

1. Write in index form.
   (a) the fifth root of n (b) $\sqrt[4]{n^3}$

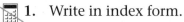 Give the answers as whole numbers or fractions.

2. (a) 5^{-1} (b) 5^0 (c) $25^{\frac{1}{2}}$ (d) $8^{\frac{2}{3}}$ (e) $27^{\frac{2}{3}}$

3. (a) 3^{-2} (b) $\left(\frac{1}{4}\right)^{-1}$ (c) $100^{\frac{5}{2}}$ (d) $10\,000^{\frac{3}{4}}$ (e) $32^{\frac{3}{5}}$

4. (a) $3^2 \times 16^{\frac{1}{2}}$ (b) $3^{-1} \times 5^{-2}$ (c) $2^{-2} + 6^0 + 5^{-1}$

Give the answers exactly or correct to 5 s.f.

5. (a) 1.47^5 (b) 0.13^3 (c) 1.15^{10} (d) 4.7^{-3}

6. (a) $1\,771\,561^{\frac{1}{6}}$ (b) $25.72^{\frac{1}{4}}$ (c) $15^{\frac{4}{5}}$ (d) $\sqrt[5]{10}$

Exercise 22.2c

1. Write as powers of 2 as simply as possible.
   (a) 64 (b) $16^{\frac{3}{4}}$ (c) 0.125 (d) 1 (e) $2^3 \times \sqrt{2}$ (f) $2^n \times 2^{n+1}$

2. Write as a power of 2 and 3 as simply as possible.
   (a) 72 (b) $\frac{9}{16}$ (c) $\sqrt{6}$ (d) $12^{\frac{2}{3}}$ (e) 18^{3n}

3. Simplify where possible.
   (a) $3x^2y \times 5xy^3$ (b) $28x^3y^2 \div 4xy^2$ (c) $15xy^3 + 3x^2y$ (d) $5x^3 \times 4x^{-1}$

4. Simplify.
   (a) $(2x^3)^2$ (b) $(27x^3)^{\frac{1}{3}}$ (c) $(5x)^{-2}$ (d) $\sqrt[3]{(8x^3y^6)}$

5. Simplify.
   (a) $\dfrac{2xyz \times 3xy^2z^2}{xy^2z}$ (b) $3x(2xy - y^2) + 2y(5x^2 - 2xy)$

6. Solve the following equations.
   (a) $3^x = 81$ (b) $2^x = 0.25$ (c) $5^x = 1$ (d) $2^{x+3} = 64$.

Exponential growth and decay

1. £3000 is invested at 4% compound interest.
    (a) Calculate the value of the investment after
        (i)   2 years   (ii)   20 years.
    (b) Find a formula for the amount the investment is worth after n years.

2. A car costs £12 000 when new. It depreciates in value by 13% per year.
    (a) Calculate the value of the car after
        (i)   3 years   (ii)   8 years.
    (b) Find a formula for the value of the car after n years.

3. A colony of bacteria is found to increase by 20% every hour.
    (a) If there are 500 000 bacteria at noon, find the number at
        (i)   3 p.m.   (ii)   6.30 p.m.
    (b) Find a formula for the number after n hours.

4. Plot a graph of $y = 2.5^x$ for values of x from -2 to 4. Use a scale of 2 cm to 1 unit on the x-axis and 2 cm to 10 units on the y-axis.

    Use your graph to estimate
    (a) the value of y when $x = 3.4$
    (b) the solution to the equation $2.5^x = 11$.

Arcs and sectors

Arcs and sectors

Exercise 23.1c

1. Calculate the arc length of these sectors. Give your answers to 3 significant figures.

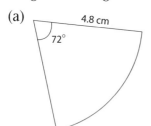

(a) 4.8 cm, 72°

(b) 7.8 cm, 304°

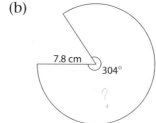

(c) 156°, 9.5 cm

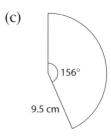

2. Calculate the areas of the sectors in Question 1. Give your answers to 3 significant figures.

3. Calculate the perimeters of these sectors. Give your answers to 3 significant figures.

(a) 5.2 cm, 100°

(b) 17°, 8.4 cm

4. Calculate the sector angle in each of these sectors. Give your answers to the nearest degree.

(a) 5.6 cm, 5.6 cm

(b) 43.4 cm, 10.2 cm

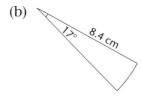

(c) 3.8 cm, Area = 8.2 cm²

(d) Area = 50 cm², 7.3 cm

5. Calculate the radius of each of these sectors.

(a)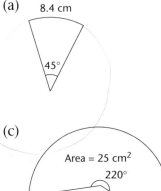
8.4 cm
45°

(b)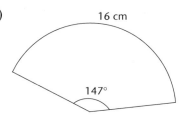
16 cm
147°

(c)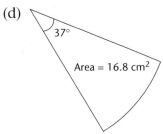
Area = 25 cm²
220°

(d)
37°
Area = 16.8 cm²

Volumes and surface areas

Exercise 23.2c

1. Calculate the volumes of these pyramids. Their bases are squares or rectangles.

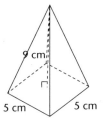
9 cm
5 cm   5 cm

5 cm
4 cm   6 cm

2. Calculate the volumes of these cones.

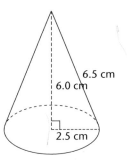
6.5 cm
6.0 cm
2.5 cm

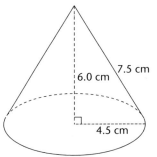
7.5 cm
6.0 cm
4.5 cm

3. Find the volume to 3 s.f of a sphere of radius

   (a) 4.6 cm

   (b) 8 mm.

4. Calculate the curved surface area of these cylinders

   (a) radius = 6.1 cm, height = 8.9 cm

   (b) radius = 4 mm, length = 18 mm.

5. Calculate the curved surface area of the cones in Question 2.

6. Calculate the curved surface area of the spheres in Question 3.

7. This cylindrical jug holds 2 litres [2000 cm^3].
   Its base radius is 8 cm. Find its height.

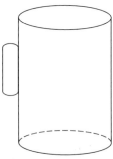

8. This conical glass holds 150 ml.
   Its top radius is 4 cm. Find its depth, d.

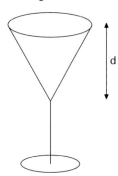

More complex problems

1.

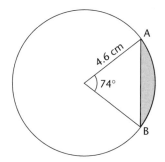

Calculate
(a) the length of the chord AB
(b) the perimeter of the shaded segment.

2.

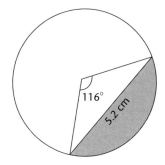

Calculate the area of the shaded segment.

3.

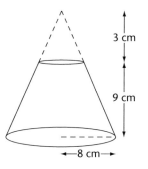

A cone of height 12 cm and base radius 8 cm
has the top 3 cm removed.
(a) Find the radius of the top
     remaining frustrum.
(b) Calculate the volume of the frustrum.

4.

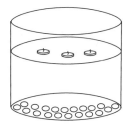

A cylindrical glass bowl of radius 15 cm has
water in it with floating candles. 20 blue glass
spherical marbles of radius 1.2 cm are placed in
the bowl. By how much does the water level in
the bowl increase?

5. A cone has base radius 4.3 cm and height 8.4 cm. It has the same
volume as a sphere. Find the radius of the sphere.

Chapter 23   Exercise 23.3   Exercise 23.3

Volumes and surface areas of similar figures

Exercise 23.4c

1. State the area scale factor for these length scale factors:
   (a) 4   (b) 1.5.

2. State the volume scale factor for these length scale factors:
   (a) 6   (b) 20.

3. State the length scale factor for the following:
   (a) area scale factor of 36
   (b) volume scale factor of 8.

4. An oval mirror has an area of 162 cm^2. What is the area of a similar mirror one and a half times as long?

5. The three tables in a nesting set of tables are similar, with heights in the ratio 1 : 1.2 : 1.5. The area of the smallest table top is 120 cm^2. What is the area of the middle sized table top?

6. A model of a theatre set is made to a scale of 1 : 20. A cupboard on the model has a volume of 50 cm^3. Find the volume of the cupboard on the actual set, giving your answer in cubic metres.

7. Two jugs are similar and contain 1 litre and 2 litres respectively. The height of the larger jug is 14.8 cm. What is the height of the smaller jug to the nearest millimetre?

Upper and lower bounds

Sums and differences of measurements

Exercise 24.1c

1. Calculate the upper bounds of the sums of these measurements.
   - (a) 43.2 cm and 81.7 cm (both to the nearest millimetre)
   - (b) 10.31 s and 19.17 s (both to the nearest $\frac{1}{100}$ second)
2. Find the lower bounds of the sums of the measurements in Question 1.
3. Find the upper bounds of the differences of these measurements.
   - (a) 489 m and 526 m (both to the nearest metre)
   - (b) 0.728 kg and 1.026 kg (both to the nearest gram)
4. Find the lower bounds of the differences of the measurements in Question 3.

Multiplying and dividing measurements

Exercise 24.2c

1. Two people measure a room. Find the upper bounds for the area of the floor for both measurements.
   - (a) 4.3 m × 6.2 m, both to 2 s.f.
   - (b) 4.27 m × 6.24 m, both to the nearest centimetre.
2. Find the lower bounds for the area of the floor in Question 1.
3. Calculate the upper bounds for the average speeds. Give your answers to 4 s.f.
   - (a) 157 km (nearest km) in 2.5 hours (nearest 0.1 hour)
   - (b) 800 cm (nearest 10 cm) in 103.47 s (nearest $\frac{1}{100}$ s)
4. Find the lower bounds for the average speeds in Question 3.

Problems in 3D and non-right-angled triangles

Section A : 3D problems

1.

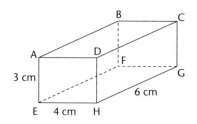

ABCDEFGH is a rectangular box with dimensions as shown. Calculate:

(a) AC

(b) angle BAC

(c) AG

(d) angle CAG.

2.

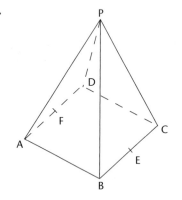

PABCD is a square-based pyramid with P vertically above the midpoint of the square ABCD. Given that AB = 12 cm, AP = BP = CP = DP = 15 cm, E is the midpoint of BC and F is the midpoint of AD, calculate:

(a) DB

(b) angle PBD

(c) PE

(d) angle PEF.

3.

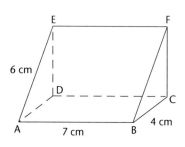

ABCDEF is a triangular prism, with angle BCF = 90°. Calculate:

(a) FC

(b) angle FBC

(c) EB

(d) angle EBD.

The angle between a line and a plane

Exercise 25.2c

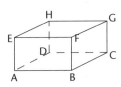

ABCDEFGH is a rectangular box.

1. Sketch the triangle and label the angle between the following lines and planes:

    (a) AG and ABCD   (b) EC and BCGF.

2. Given that AB = 12 cm, BC = 8 cm and BF = 9 cm, calculate the angles between the following lines and planes:

    (a) AG and ABCD   (b) EC and BCGF.

Section B: non-right-angled triangles and the Sine Rule

Exercise 25.3c

1.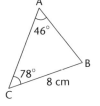

    Find c, B and b.

2.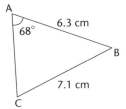

    Find C, B and b.

3. In the triangle EFG, EF = 7 cm, FG = 9 cm and angle G = 39°. Find:
    (a) angle E   (b) angle F   (c) side EG.

4. In the triangle PQR, PQ = 7.8 cm, angle R = 79°, angle P = 51°. Find:
    (a) side QR   (b) angle Q   (c) side PR.

5.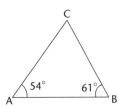

    ABC is a flower bed in a garden. AB is 3.56 m long. Find the length of the other two sides.

The Cosine Rule

1. 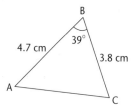 Find length AC.

2. 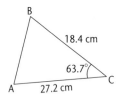 Find length AB.

3. 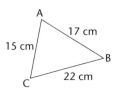 Find angle CAB.

4. 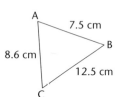 Find angle CAB.

5. Find
   (a) BD
   (b) AB.

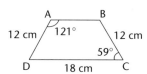

General formula for the area of any triangle

Exercise 25.5c

1. Find the area of each of the following triangles.

    (a)

    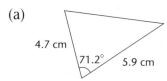

    (b)

    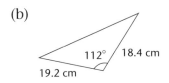

2. In triangle ABC, AB = 15.3 cm, AC = 9.6 cm and angle BAC = 53.6°.
    Find the area of triangle ABC.

❦ Conditional probability

Conditional probability

1. There are 8 boys and 12 girls in a class. Two students are selected at random from the class.

   (a) Draw a tree diagram to show the probabilities of the possible outcomes.

   (b) Find the probability that

   (i)   two girls are selected

   (ii) one boy and one girl is selected.

2. A hand of 13 cards contains 6 red cards and 7 black cards.

   Two cards are selected from the hand at random.

   (a) Draw a tree diagram to show the probabilities of the possible outcomes.

   (b) Find the probability that

   (i)   both cards selected are black

   (ii) at least one card selected is red.

3. The probability that a plum tree will produce more than 50 kg of plums in any year is 0.6.

   If it produces more than 50 kg of plums one year, then the probability that it will produce more than 50 kg of plums in the following year is 0.8, if it does not then it is 0.4.

   (a) Draw a tree diagram to show the probabilities of the possible outcomes.

   (b) Find the probability that the tree will

   (i)   produce more than 50 kg of plums for two consecutive years

   (ii) produce more than 50 kg of plums on one of the two consecutive years.

4. A box contains 4 red, 5 blue and 6 green counters.

    Two counters are chosen at random without replacement.

    Find the probability of

    (a) choosing two counters of the same colour

    (b) choosing one blue counter.

5. On the way home I pass through two sets of traffic lights.

    The probability that the first set is on green is 0.5.

    If the first set is on green,

    the probability that the second set is on green is 0.6.

    If the first set is not on green.

    the probability that the second set is on green is 0.3.

    Find the probability that

    (a) I do not have to stop at either of the sets of traffic lights

    (b) I have to stop at at least one set of traffic lights

    (c) I have to stop at both sets of traffic lights.

27 Algebraic laws

General laws in symbolic form and sequences

Exercise 27.1c

In questions where you choose your own letters make sure you define them carefully.

1. Coaches carry 50 passengers and double-decker buses carry 80 passengers.

   (a) Find a formula for the total number of passengers carried by coaches and double-decker buses.

   (b) If it is necessary to carry 500 passengers, write down an inequality satisfied by the number of coaches and double-decker buses needed.

2. Exercise books cost 40p and text books cost £12. Find a formula for the total cost of buying exercise books and text books.

3. A semicircle is cut from a rectangle as shown in the diagram.

   Find a formula for the area remaining.

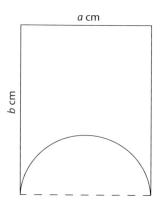

a cm

b cm

4. The table shows a frequency distribution.

x	Frequency
a	p
b	q
c	r

   Write down a formula for the mean value of x.

5. Brian cycles to see his friend who lives a distance of d km away. He cycles there at x km/h and returns at y km/h.

    (a) Find a formula for the total time taken for the complete journey.

    (b) Find a formula in terms of d, x and y for the average speed for the complete journey.

Approximation to linear graphs and non-linear graphs

Exercise 27.2c

1. In an experiment the quantities x and y are measured. The table of values is given below.

x	1	2	3	4	5
y	7.1	30.1	61.3	115	183

    (a) Draw a graph of y against x.

    (b) It is thought that the relationship between x and y is of the form $y = ax^2$.

    Confirm this by plotting a graph of y against x^2.

    (c) Draw a line of best fit and use it to estimate the value of a.

2. In an experiment the quantities x and y are measured. The table of values is given below.

x	10	20	30	40	50
y	43.9	21.8	15.3	11.2	9.1

    (a) Draw a graph of y against x.

    (b) It is thought that the relationship between x and y is of the form

    $$y = \frac{a}{x}.$$

    Confirm this by plotting a graph of y against $\frac{1}{x}$.

    (c) Draw a line of best fit and use it to estimate the value of a.

Circle properties

Angles at the centre of a circle

Exercise 28.1c

In each of the following diagrams, O is the centre of the circle. Find the size of each of the lettered angles. Write down the reasons for your deductions.

1.
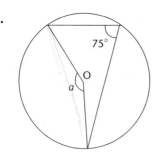

2.
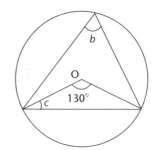

3.
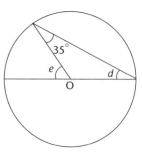

Exercise 28.2c

In the following questions O is the centre of the circle. Calculate the angles marked with letters.

1.
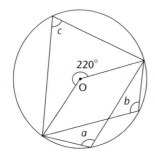

2.
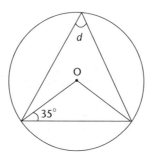

3.
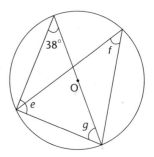

Calculate the sizes of the angles marked with letters. O is the centre of each circle. Give the reasons for each step of your working.

1.
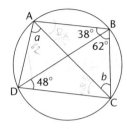

2.
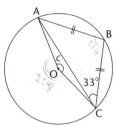

3.
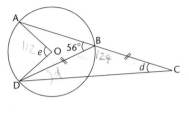

Tangents

In the following questions calculate the angles marked with letters. O is the centre of each circle. X and Y are the points of contact of the tangents to each circle.

1.
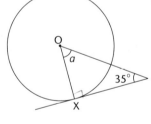

2.
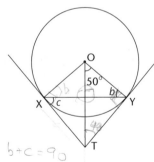

3.
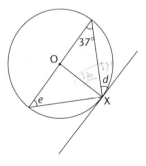

In these questions TX is a tangent to the circle. Find the angles marked with letters. Give reasons for your answers.

1.
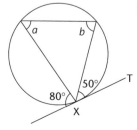

2.
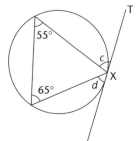

3.
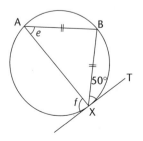

29 Equations and manipulation 4

Revision of earlier algebra

Exercise 29.1c

1. Solve by factorisation.

   (a) $x^2 - 5x - 24 = 0$ 　　　　　(b) $2x^2 - 9x + 9 = 0$

2. Use the formula to solve the following. Give the answers to 2 d.p.

   (a) $x^2 - 3x + 1 = 0$ 　　　　　(b) $3x^2 + x - 5 = 0$

3. Simplify.

   (a) $\dfrac{1}{x} - \dfrac{3}{x-1}$ 　　　　　(b) $\dfrac{2x}{x+1} + \dfrac{3x}{x+3}$

4. Solve the equations.

   (a) $\dfrac{5x+5}{2x-1} = x + 1$ 　　　　　(b) $\dfrac{2}{x+1} + \dfrac{3}{x-2} = 1$

5. Rearrange the formula to make a the subject.

   (a) $b = \dfrac{a+1}{a-1}$ 　　　　　(b) $2(a + b) = 3(c - a)$

6. Solve the simultaneous equations.

   (a) $3x + 2y = -5$ 　　　　　(b) $5x + 2y = 7$
   　　$x - 4y = -4$ 　　　　　　　$2x - 3y = 18$

Further simultaneous equations

Exercise 29.2c

Solve the following simultaneous equations by the method of substitution.

1. $2x + 3y = -5$ 　　　2. $x + y = 2$ 　　　3. $y + x - 3 = 0$
   $x - y = 10$ 　　　　　　$3x + 2y = 2$ 　　　　$y = x^2 + 1$

4. $y = x^2 + x$ 　　　　5. $y = x^2 + 3x - 1$
   $y = x + 1$ 　　　　　　$x - 2y - 4 = 0$

The equation of a circle

1. (a) Draw the graphs of $x^2 + y^2 = 25$ and $y = x + 1$ on the same grid.
   (b) Use the graph to solve simultaneously the two equations.

Use algebra to solve the following equations simultaneously.

2. $x^2 + y^2 = 2$ $y = 2x - 1$
3. $x^2 + y^2 = 36$ $x + y = 5$
4. $x^2 + y^2 = 9$ $2y - x = 1$

Gradients of parallel and perpendicular lines

1. Find the gradient of the lines joining the points
   (a) (1, 1) and (7, 3)   (b) $(-2, 1)$ and $(3, -6)$.

2. Find the equation of the lines through the points
   (a) (2, 1) and (6, 4)   (b) $(-2, -2)$ and $(-4, 6)$.

3. Find the equation of the line that is parallel to the line $y = 2x - 1$ and passes through the point $(-3, -1)$.

4. Find the equation of the line that is perpendicular to the line $2y + x = 0$ and passes through the point $(6, -2)$.

5. Find which of these lines are (a) parallel or (b) perpendicular to $y = 2x + 1$.
   $2y = x + 1$ $2y - 4x = 3$ $2y + x = 1$ $2y + 4x = 3$

Histograms and standard deviation

Histograms

1. The heights of 120 children are measured. The results are shown in this table.

   Draw a histogram to show this information

Height	Number of Children
$75 \leqslant h < 100$	15
$100 \leqslant h < 120$	20
$120 \leqslant h < 140$	32
$140 \leqslant h < 160$	44
$160 \leqslant h < 180$	9

2. The number of men and women in each age group at the Senfield Golf club is calculated. The table shows the results.

Age (years)	Men	Women
15–19	4	2
20–29	10	8
30–39	18	12
40–49	12	14
50–64	32	25
65–79	45	12

   (a) Show the information for men and for women on two separate histograms.

   (b) Compare the age distributions.

1. This histogram shows the distribution of time spent watching TV in a week by a group of people.

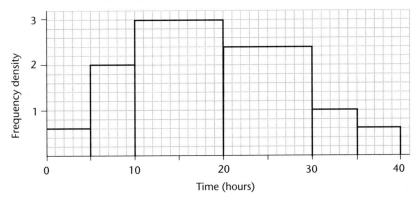

There were three people who watched TV for 0 to 5 hours each week.

(a) How many were there in each other group?

(b) Work out the mean length of time spent watching TV.

2. This histrogram shows the age distribution of the members of Carterknowle Methodist Church.

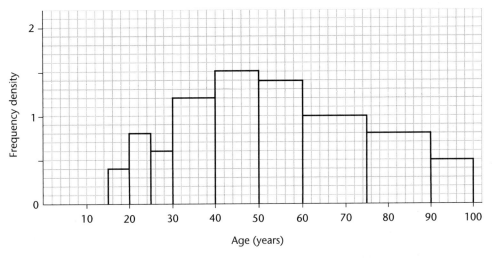

There are 2 members under 20.

(a) How many members are there altogether?

(b) Work out the mean age of the members.

31 Trigonometrical functions

Trigonometrical functions of any angle

Exercise 31.1c

1. Sketch the graph of $y = \sin x°$ for values of x from 0 to 360. Use your graph to find the solutions of $\sin x° = 0.7071$.

2. Sketch the graph of $y = \cos x°$ for values of x from -360 to 360. Use your graph to find the solutions of $\cos x° = 0.866$.

3. Sketch the graph of $y = \sin x°$ for values of x from -180 to 360. Use your graph to find the solutions of $\sin x° = -0.788$.

4. Given that $\cos 40° = 0.766$, use the symmetry of the graph of $y = \cos x°$ to find the solutions of $\cos x° = -0.766$ for x between 0 and 360.

Other trigonometrical graphs

Exercise 31.2c

1. Draw accurately the graph of $y = 4\sin x°$ for values of x from 0 to 360 plotting values every 30.

2. Draw accurately the graph of $y = \cos 4x°$ for values of x from 0 to 180 plotting values every 10. State the period and the amplitude of the graph.

3. Sketch the graph of $y = \cos \frac{1}{2}x°$ for values of x from 0 to 360.

4. Find the solutions of $5\sin x° = 1$ for x between 0 to 360.

5. Find the solutions of $\cos 4x° = 1$ for x between 0 to 360.

Rational and 32 irrational numbers

Rational and irrational numbers and recurring decimals and fractions

Exercise 32.1c

1. State whether each of these numbers is rational or irrational, showing how you decide.

   (a) $\dfrac{5\pi}{2}$ (b) $0.5\dot{4}$ (c) $\sqrt{144}$ (d) $\sqrt{66}$ (e) $5 + 2\sqrt{3}$

2. Convert these fractions to recurring decimals using the dot notation.

   (a) $\dfrac{7}{9}$ (b) $\dfrac{5}{18}$ (c) $\dfrac{17}{303}$

3. Convert these recurring decimals to fractions in their lowest terms.

   (a) $0.\dot{7}\dot{2}$ (b) $0.4\dot{8}$ (c) $0.\dot{3}0\dot{6}$

Simplifying surds and rationalising denominators

Exercise 32.2c

1. Simplify the following, stating whether the result is rational or irrational.

   (a) $\sqrt{28}$ (b) $\sqrt{63}$ (c) $\sqrt{125}$ (d) $\sqrt{600}$ (e) $\sqrt{8} \times \sqrt{10}$

2. If $x = 5 + \sqrt{2}$ and $y = 5 - \sqrt{2}$, simplify

   (a) $x + y$ (b) $x - y$ (c) xy.

3. If $x = 7 + \sqrt{3}$ and $y = 5 - 2\sqrt{3}$, simplify

   (a) $x + y$ (b) $x - y$ (c) x^2 (d) xy.

4. Rationalise the denominator in these irrational fractions

   (a) $\dfrac{1}{\sqrt{3}}$ (b) $\dfrac{3}{\sqrt{5}}$ (c) $\dfrac{7}{\sqrt{10}}$ (d) $\dfrac{10}{\sqrt{5}}$ (e) $\dfrac{5}{2\sqrt{3}}$.

Transformations and functions

Function notation

Exercise 33.1c

1. If $f(x) = 4x - 3$, find the values of
   (a) $f(6)$ (b) $f(-2)$.
2. If $g(x) = x^2 - 7x + 6$
   (a) find the values of   (i) $g(0)$ (ii) $g(5)$
   (b) solve by factorisation the equation $g(x) = 0$.
3. If $h(x) = 2x + 5$, write expressions for
   (a) $h(3x)$ (b) $2h(x)$ (c) $h(x) - 4$.

Translations

Exercise 33.2c

1. Sketch on the same diagram the graphs of (a) $y = x^2$ and (b) $y = x^2 - 4$.
   State the transformation which maps (a) onto (b).
2. Sketch on the same diagram the graphs of (a) $y = x^2$ and (b) $y = (x - 3)^2$.
   State the transformation which maps (a) onto (b).
3. State the equation of the graph of $y = x^3$ after it has been
   translated by $\begin{pmatrix} -2 \\ 0 \end{pmatrix}$.
4. State the transformation which map the graph of $y = x^2$ onto the graph of
   $y = (x - 5)^2 + 6$.
5. 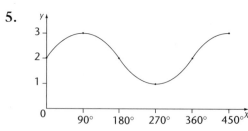 This graph is a transformed sine curve. State its equation.

One-way stretches and reflections

Exercise 33.3c

1. Sketch on the same diagram the graphs of $y = \sin x°$ and $y = 2\sin x°$ for $0 \leqslant x \leqslant 360$. Describe the transformation that maps $y = \sin x°$ onto $y = 2\sin x°$.

2. Describe the transformation that maps $y = \cos x°$ onto $y = \cos \frac{1}{2}x°$.

3. Sketch on the same diagram the graphs of $y = \cos x°$ and $y = \cos 4x°$ for $0 \leqslant x \leqslant 180$. Describe the transformation which maps $y = \cos x°$ onto $y = \cos 4x°$.

4. The graph of $y = x^3 + 5$ is reflected in the x-axis. State the equation of the resulting graph.

5. The graph of $y = x^3 + 2$ is reflected in the y-axis. State the equation of the resulting graph.

6. State the equation of the graph $y = x^3 + 2$ after it is stretched:

   (a) parallel to the y-axis with scale factor 3

   (b) parallel to the x-axis with scale factor $\frac{1}{4}$.

Transforming relationships to a linear form

Exercise 33.4c

1. State the variables you would need to plot to obtain a straight-line graph of the following equations:

   (a) $V = \dfrac{10}{R}$ (b) $S = 6 + 8t^2$ (c) $C = 4\sqrt{h} - 1$.

2. Use these graphs to find the equations connecting the variables.

   (a)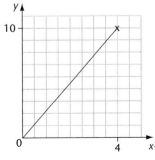

   (b)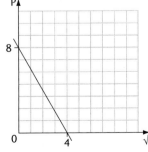

3. (a) Complete the missing row in this table of values.

t	1	2	4	5	10
$\frac{1}{t}$					
y	14	8	5	4.4	3.2

(b) Plot a graph of y against $\frac{1}{t}$ and hence find the equation connecting y and t.

Using functions to model problems

1. (a) Complete this table for the function $y = 4^x$.

x	0	1	2	3	4	5
y						

(b) Find the value of y when $x = 10$.

(c) Using trial and improvement, or otherwise, find correct to 1 decimal place the value of x when $y = 10\,000$.

2. A population of butterflies is declining at 8% a year. The population in August 2001 was 850.

(a) Explain why a suitable equation for the population t years later is given by $y = 850 \times 0.92^t$.

(b) Find the population in August 2006.

(c) In August of which year is the population first below 400?

3. The depth, d m, of water in a harbour at time t hours is given by

$d = 5 + 2 \sin(30t)°$.

(a) Sketch a graph of this function for $t = 0$ to 12.

(b) At what time is low water, and what is the depth of water in the harbour then?

(c) A ship can enter the harbour when the water in the harbour is more than 6 m deep. Find the range of values of t during the period $t = 0$ to 12 for which the ship can enter the harbour.

4. The curve $y = ab^x$ passes through the points (0, 10) and (4, 810). Find the valves of a and b.

5.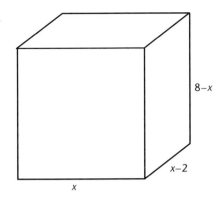

The diagram shows the dimensions, in centimetres, of a cuboid.

(a) Write a formula for V in terms of x where $V\,\mathrm{cm}^3$ is the volume of the cuboid.

(b) Draw a graph of V against x for $x = 2$ to 8.

(c) Find the dimensions, correct to the nearest millimetre, which give the maximum volume for the cuboid.

34 Vectors

Column vectors and translations

Exercise 34.1c

1. For the diagram below write down the vectors $\overrightarrow{PQ}$, $\overrightarrow{RS}$, $\overrightarrow{TU}$, $\overrightarrow{VW}$, and $\overrightarrow{XY}$ in terms of **a** and **b**.

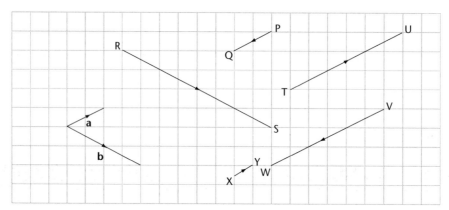

2. For the diagram below write down the column vectors for $\overrightarrow{AB}$, $\overrightarrow{AD}$, $\overrightarrow{CB}$, and $\overrightarrow{DC}$.

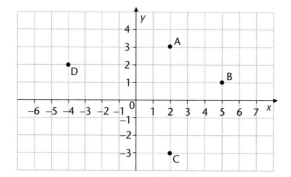

3. Describe fully the transformation that maps
   (a) A onto B   (b) C onto D   (c) D onto B   (d) A onto D.

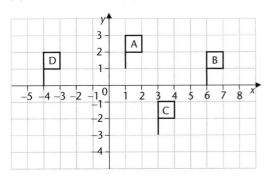

4. Find the column vector that maps
   (a) (1, 3) to (2, 5)   (b) (−2, 0) to (3, −1)   (c) (−5, 2) to (−5, −5)
   (d) (3, 4) to (0, 0).

Combining column vectors

Exercise 34.2c

1. Work out   (a) $3 \times \begin{pmatrix} 1 \\ 2 \end{pmatrix}$ (b) $\begin{pmatrix} 3 \\ 4 \end{pmatrix} + \begin{pmatrix} 1 \\ 6 \end{pmatrix}$ (c) $\begin{pmatrix} 3 \\ 4 \end{pmatrix} - \begin{pmatrix} 1 \\ 6 \end{pmatrix}$.

2. Given that $\mathbf{a} = \begin{pmatrix} -3 \\ 0 \end{pmatrix}$, work out

   (a) $3\mathbf{a}$ (b) $-\mathbf{a}$ (c) $-\frac{1}{3}\mathbf{a}$.

3. Given that $\mathbf{a} = \begin{pmatrix} 1 \\ 2 \end{pmatrix}$ and $\mathbf{b} = \begin{pmatrix} 1 \\ 3 \end{pmatrix}$, work out

   (a) $\mathbf{a} - \mathbf{b}$ (b) $2\mathbf{a} + 3\mathbf{b}$ (c) $3\mathbf{a} - 2\mathbf{b}$.

4. Given that $\mathbf{a} = \begin{pmatrix} 2 \\ 0 \end{pmatrix}$, $\mathbf{b} = \begin{pmatrix} -3 \\ 1 \end{pmatrix}$ and $\mathbf{c} = \begin{pmatrix} 0 \\ -4 \end{pmatrix}$, work out

   (a) $\mathbf{a} + \mathbf{b} - \mathbf{c}$ (b) $2\mathbf{a} - \mathbf{b} + 3\mathbf{c}$ (c) $\frac{1}{2}\mathbf{a} + \frac{1}{2}\mathbf{b} + \frac{1}{2}\mathbf{c}$.

Vector geometry

Exercise 34.3c

1. The vectors **a** and **b** are drawn on the grid.

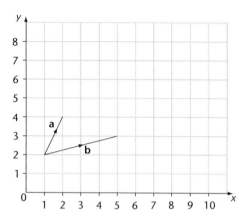

   Draw the resultant of 2**a** + 2**b**.

2.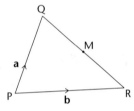

   (a) Work out the vector $\overrightarrow{QR}$ in terms of **a** and **b**.

   (b) M is the midpoint of QR. Find the vector $\overrightarrow{PM}$ in terms of **a** and **b**.

3. PQRS is a parallelogram. The midpoints of PQ, QR, RS and PS are I, J, K and L respectively.
   $\overrightarrow{SR}$ = **a** and $\overrightarrow{RQ}$ = **b**.

   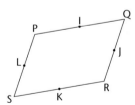

   Find in terms of **a** and **b**
   (a) $\overrightarrow{PQ}$ (b) $\overrightarrow{PS}$ (c) $\overrightarrow{SQ}$ (d) $\overrightarrow{LI}$.
   (e) What do the vectors show about LI and SQ?

4. ABCD is a rectangle. The midpoints of AB, BC, CD and DA are E, F, G, and H respectively.

$\overrightarrow{AB} = \mathbf{p}$ and $\overrightarrow{BC} = \mathbf{q}$.

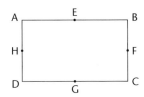

Find in terms of **p** and **q** (a) $\overrightarrow{EF}$ (b) $\overrightarrow{HG}$.

(c) What does this tell you about EFGH?

5. Triangle OAB is enlarged, centre O, scale factor -2 to give OCD.

$\overrightarrow{OA} = \mathbf{a}$ and $\overrightarrow{OB} = \mathbf{b}$.

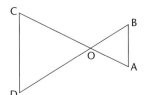

Use vectors to show that CD is parallel to BA.